EXECUTING IT PROJECT
MANAGEMENT WITH SOFTWARE
QUALITY ASSURANCE AND TESTING

A GUIDE FOR CIOs AND SENIOR MANAGEMENT

JOHN J. SCARPINO, D.Sc., MBA

BIOGRAPHY

I have primarily been working in Information Technology (IT) and Software Development for over twenty-five years. In the beginning of my career, I started as a network IT engineer, then as a Software Engineer, Sr. Software Engineer, Manager, Director, Assistant Vice President, and then Vice President within the fields of IT Networking, Software Testing, Software Quality Assurance (QA), QA Center of Excellence, Project Management, Project Management Office, and Program Management Offices (PMO). I have experience across multiple industries, such as healthcare, health insurance, surgical transplant software, telecommunications, education services, finance, retail, global consulting, and aerospace and defense. Working in surgical software was especially meaningful, as our team's efforts supported organ transplants and positively affected patients, clinicians, and doctors. Leading high-quality teams in this field was both impactful and rewarding.

At one point in my career, I was able to work at a small college leading IT and running academics. I began as Chair for Information Systems and Technology before becoming the founding Dean of the School of Engineering, Technology and Trades, overseeing departments like IT, Cyber Security, Networking, Electronics, Engineering and Design, CAD, Media Arts and Communications, Welding, and HVAC. During the college's peak, I contributed to expanding technology, enrollment, programs, grant funding, and establishing its first IRB. These efforts helped grow the college until its closure in June 2024, alongside many other institutions after the post-COVID era.

I co-owned a cybersecurity company for two years before selling my 50% stake, and I also solely owned a software development consulting firm. During all this, I enjoyed giving back to the community by teaching undergrad and graduate classes as an adjunct professor from

2002 to 2023 at Robert Morris University. Recently in 2020 during the pandemic, I have taught in Pullman's MBA program (specifically teaching Management Information Systems) as an adjunct online at Washington State University. During these times, I have taught and put together graduate and undergraduate classes and degree programs in Networking, Hardware, Software Development, Project Management, Cybersecurity, Quality Assurance, Systems Analysis, Data Analytics, Artificial Intelligence, Decision Support, Database, Web, and Mobile Development. My industry specialty and research repertoire are all around software QA and testing, portfolio and project management, business analysis and requirements traceability, software and infrastructure quality assurance ensuring its development, security, assurance, SDLC process, implementation and planning.

I have a BSBA in Finance and double majored in Management Information Systems, a Leadership Certificate from Harvard University, a Master of Science (MS) in Internet Information Systems from Robert Morris University, an MBA with a concentration in Management Information Systems from Point Park University, and a Doctor of Science (D.Sc.) in Information Systems and Communications also from Robert Morris University's, School of Data Intelligence and Technology. My dissertation "An Analysis of a Post-Implementation Survey of Users Perceptions toward an Implementation of Software Quality Assurance Testing Tools (SQATT) at a Fortune 500 Financial Institution" was completed at Robert Morris University in 2010 and published in UMI / ISBN:978-1-124-45288-3.

In 2014, I was inducted as an honorary member of the Alpha Iota Mu. Alpha Iota Mu (AIM) is the honor society specifically targeted to Information Systems majors and minors. The award is limited to those people of high scholarship and good moral character who strive for excellence in the field of Information Systems Management, Information Science, or any Information Systems-related area. This was a great honor as I was able to share this time with my late mother.

I also have over thirty research publications and proceedings with several being awarded or recognized for best paper. Feel free to visit my website for more information: www.johnscarpino.com or www.softwarequalityanalyst.com

ACKNOWLEDGEMENTS

To my late mother and my elder father, I owe both life itself and life's learning too. Thank you for the sacrifices you made to help me grow, learn, and have a positive impact on others. May God bless you both. My mother would say while she was in the hospital: fight, eat, live, breathe, and pray, every day as she strongly fought and won! If we could all remember doing this every day like my mom, we would live stronger and with more meaning. I have been able to help others because of your gifts to me.

To my wife for her support. I have been working on this book for close to fifteen years since I graduated with my doctorate (hard to put aside a career and family for this work). Thank you for your love, support can partnership. We both support each other's careers, and I am thankful that you support my goals.

To my beautiful, smart, and growing loving daughter, may you continue to grow and be an inspiration to others as you get older. I hope that STEAM (Science, Technology, Engineering, Arts, and Math) can be part of your future, and I can already see its influence on you. Always lead by example and be among those who make a difference.

To my late doctoral mentor Dr. Anthony "Tony" Debons from the University of Pittsburgh (EATPUT model), I miss the late evenings having dinner with you at your house. Parts of this book were reviewed with him many years ago in his living room, sunroom and home office. Thank you for your inspiration, lessons, and growth. I will forever cherish those times.

To my former professors, doctorate and graduate advisors, faculty, colleagues, team members and friends from Robert Morris University, Washington State University, and the former Pittsburgh Technical

College, thank you for the twenty-plus years of work in academic, research, and life's inspiration. All your assistance and experiences helped me grow in both academics, research, and within the industry.

Then, finally, to all my friends, industry colleagues, and team members from the software engineering and IT industry, I have countless amounts of gratitude, respect, and thanks for you. We felt pain together, grew and learned together, and put out some great technical projects, deployments, and late-night implementations and fixes. Again, thank you all!!

EXECUTIVE SUMMARY

The purpose of this book is to help companies and managers apply proper quality practices within their management of programs and projects, help the software quality assurance or software testing teams, and assist scholars researching this field. I hope that these tips and strategies help you implement better techniques of influence for better quality results.

Executing IT Project Management with Software Quality Assurance and Testing: A Guide for CIOs and Senior Management

CONTENTS

INTRODUCTION

The fields of IT project management, software quality assurance, and software testing have grown tremendously over the last twenty-plus years. Though testing has always existed since software development first started, the formality of software testing, from my perspective, began when multiple browsers existed with multiple versions and operating systems creating complications. Plus, with the invention of the GUI interfaces, this created a compounding effect in more than one way of how issues and defects existed. Now, it's much more than simply finding a bug or a defect but how can management support and understand how to ensure its quality and excellence for future growth and assurance. Owners, CIOs, and executive managers all need to be on the same page when it comes to formulating a process that fits the organization's culture and understanding how to create the right implementation of quality assurance and its software testing. It's more than just hiring people or implementing software but rather how to intentionally implement the right management, processes, and techniques to ensure and foster ever-growing risks and quality.

I hope that this book helps to shed some light on this topic and helps the management of software quality assurance and its testing. In 2024 and 2025, we have seen multiple quality concerns. For one, NASA had doubts of being able to bring back two astronauts from space due to Boeing's defect issues. The two astronauts were then eventually brought back by SpaceX in 2025. Then there was the 2024 CrowdStrike defected update, which impacted hospitals, supply chains, and the airline industry. All of this could have been prevented with better methods, practices, and processes in engineering and testing. There is an opportunity to take on all these quality issues and understand that there should be stronger quality practices, processes, and testing to help ensure better products and services and to reduce risks. In the end, my hope is that this book will help you better implement projects and processes with enhanced quality.

Chapter One
QA MANAGEMENT: TIPS AND STRATEGIES

"The aim of leadership should be to improve the performance of man and machine, to improve quality, to increase output, and simultaneously to bring pride of workmanship to people. Put in a negative way, the aim of leadership is not merely to find and record failures of men, but to remove the causes of failure: to help people to do a better job with less effort."
—W. Edwards Deming

CIO EXPECTATIONS FOR QUALITY ASSURANCE: FIVE HURDLES QA MANAGERS MUST OVERCOME

Throughout my experience in software testing and software quality assurance across finance, telecommunications, healthcare, ecommerce, eMarketing, insurance, and online education, it has become clear that not all CIOs expect the same output from Quality Assurance Managers. Much of this depends on how knowledgeable the CIO is about quality assurance, the size and culture of the corporation, and whether it is a research-based institution, a traditional Fortune 500 corporation, or a start-up organization.

There are many hurdles that QA Managers must overcome in order to meet deliverables and exceed CIOs' expectations. The purpose of this piece is to shed light on some lessons I've learned from some of the best and brightest in the IT industry, particularly regarding

CIO expectations about quality assurance including what you should and should not do.

Hurdle #1: Defining Software *Quality* vs. Software *Testing*

Some CIOs, especially at large corporations, do not understand the difference between Software Quality Assurance, which ensures the usability of software quality processes and practices, and Software Testing, which carries out the actual testing activities for the application. But for some CIOs, the two are identical. This is one of the first hurdles that a Quality Manager or Director must face, demonstrating what QA is and what it isn't and making the case for why QA (in *addition* to Software Testing) is imperative for success.

Quality Assurance must be supported by executives at the top (namely, the CIO), to avoid the tunnel vision of "I don't care how you do it but just get the end product out the door *today!*" CIOs who understand the importance of QA and can actually explain his or her expectations of "good enough" will consequently have a good technical understanding of the resources needed to maintain an IT organization and make it successful.

I have seen many CIOs say that QA can be outsourced. They think from this perspective because, to them, software testers are *bodies* and not *assets*. QA is the last department you would want to outsource because it means giving up control and risks of your product to another company.

Hurdle #2: Removing Dollar Signs from the "Quality Equation"

A CIO's level of risk acceptance depends on how public facing the application and its systems are, and how many bugs the application can handle before they start to have an adverse effect on the product's profit margin. Much to the chagrin of QA Managers, one of a CIO's top priorities is to push a product to go "live" as quickly as possible for both business and personal reasons. Obviously, the sooner the product goes live, the more revenue the company will make, and the more likely it is that the company will be viewed as a success in the public eye, which in turn means that the CIO will be championed as one of the key people who got the company there. When *quality* is judged simply based on revenue, it isn't any wonder why some CIOs are willing to overlook risks for the sake of making a client happy with an ahead-of-schedule launch. Nevertheless, financial success is not always an indicator that the company has a quality product. A CIO's number one priority should be to *prevent* risk—it is in his or her best interest and in the best interest of the company for them to do so. If not, then a reputation/corporate image disaster can result which no dollar amount can save.

Hurdle #3: Mitigating Software Risk with Available QA Resources

Achieving a completely bug-free system before product launch is no picnic, and Quality Assurance Managers often struggle to adequately communicate the implications of risk at the executive level. What's worse, even if a QA Manager does reach the CIO, their concerns may fall on deaf ears. As mentioned in Hurdle #1, CIOs tend to think of Software Testing and Software Quality Assurance in the same light. CIOs instruct their QA Managers to quickly test the product until it is just good enough to pass customer inspection and get out the door. So, when QA Managers voice their grievances about quality, the CIO may refute it saying: "The customer hasn't had a problem yet, and as long as the customer is happy, that's all that matters. I'm not going to waste precious time and resources on fixing something that isn't broken."

But consider this: just because a boat can stay afloat in calm seas does not mean it is structurally sound enough to withstand a hurricane, tsunami, or rogue waves. And if the boat sinks, the crew (i.e., QA Team) goes down with the ship, including the captain (i.e., the CIO). Quality affects the entire organization, and it is not just a departmental issue. Efficiency is built with the resources and processes that are available within the company. In other words, QA must be engrained throughout the internal aspects of the organization before it can be appropriately demonstrated through an external-facing product. In successful QA organizations, quality assurance activities are not only managed by the CIO but also by the Chief Operating Officer (COO) and sometimes even the CFO/Office of Finance. This setup requires the QA Manager to report to the CIO on matters concerning IT activity and the COO/CFO on the matter of cost. This ensures an unbiased environment removed from the confines of software development so that quality can be achieved both internally and externally. The point is to have QA infiltrate the organization and persuade the Board of Directors to accept it as an imperative for success.

Hurdle #4: Heightening Security and Performance

We, as Information Technology and Information Systems experts, understand that security and performance play a huge part in how the software will impact the customer. Unfortunately, we usually don't hear about it unless the result is a negative one. There has been a massive increase of security threats and performance issues in software applications, systems, and networks in recent years. Given the current landscape, most companies feel that maintaining the integrity of sensitive data is an important part of ensuring the quality of their product or service. The financial and healthcare sectors and third-party vendor applications are particularly wary of security issues; Sarbanes Oxley (SOX), Office of the Comptroller of the Currency (OCC), Health Insurance Portability and Accountability Act (HIPAA), and the Statement on Auditing Standards (SAS 70) / SOC 2 Type 2 are great examples of standards that ensure that data and security is managed appropriately.

Daily testing, monitoring, and controlling of these environments are needed to prevent threats and ensure that new services, products, and solutions are secure and well-balanced when used publicly. CIOs must be aware that security and performance testing, quality assurance, and quality control are all unending processes. The investment of resources to find solutions is worth it. The moment that you stop investing in your QA is when hackers and other threats to your product become not only an internal issue but also a corporate reputation issue to external constituents. For instance, if your customer is a bank whose financial records or accounts are accessed illegally (hacked) because your software product failed, it will likely turn into negative publicity for the bank, but the repercussions will be even more negative for your company. Negative media coverage can severely damage or even bankrupt a company; so, using software testing to prevent defects, security and performance risks is a no-brainer if for no other reason than to avoid bad public relations!

Hurdle #5: Product Usability and Reliability

Product "usability" is rarely taken under consideration. Always ask: "How easy is our system to use?" Your product should be measured against the competition to ensure that it is the best, fastest, and *least complicated* system available. Software test usability is often overtaken by its tremendous number of system functionalities. Just because a software system or tool works does not mean it is the best. If it takes a long time for the user to figure out all of the functionalities and actually make it work, how practical is the tool? Walking the line between "cutting-edge" and "lean stability" is key for customer satisfaction, overall success, and ROI. Think about the Nintendo Wii when it was first released; senior citizens were using it for exercise because the controller has simple buttons and uses natural body movement to perform the act. If you can use a flyswatter, you can play the Wii. Now think of a traditional Nintendo controller and related games which required a combination of up, down, left, right, and pressing two or more buttons at once to perform an action. Which is easier?

CIOs find it difficult to understand how something as intangible as Software Quality Assurance can assist with improving the bottom line, and how it can move from being an idea to proving ROI. If a CIO runs the Software Quality department loosely without control or understanding of how to make it come to fruition, then SQA's growth and niche in the company will not be a success.

THE NECESSITY OF A PMO AND SOFTWARE TESTING ORGANIZATION FOR A CIO

In the dynamic world of Information Technology (IT), a Chief Information Officer (CIO) plays a pivotal role in steering the organization towards its strategic goals. To effectively manage the complexities of IT projects and ensure the delivery of high-quality products, a CIO needs the support of a Project/Program Management Office (PMO) and a Software Testing organization. I clearly remember asking my CIO of a very large Fortune 100 financial institution to announce in his next town hall that "this year is the year of Quality!" to ensure a following culture, acceptance, and backing. He did! Which assisted with a successful kickoff for the QA Center of Excellence within the enterprise's PMO.

The Need for a PMO

Strategic Alignment

A PMO ensures that all projects are aligned with the organization's strategic objectives. It provides a framework for project prioritization, ensuring that resources are allocated to projects that offer the highest value.

Standardization

A PMO brings standardization to project and program management practices across the organization. It defines the methodologies, tools, and templates to be used, ensuring consistency and efficiency in project execution.

Governance

A PMO provides governance for projects, ensuring that they are executed within the defined scope, time, and budget. It also provides visibility into project performance, enabling the CIO to make informed decisions.

The Need for a Software Testing Organization

Quality Assurance

A Software Testing organization ensures that the software products meet the defined quality standards. It conducts various tests to identify and rectify defects before the product is delivered to the client.

Risk Mitigation

By identifying defects early in the development lifecycle, a Software Testing organization helps mitigate the risks associated with software failures. This can save the organization from costly rework and potential damage to its reputation.

Compliance

A Software Testing organization ensures that the software products comply with the regulatory standards applicable to the organization's industry. This is particularly important for organizations operating in highly regulated industries such as government, healthcare and finance.

Conclusion

In conclusion, a PMO and a Software Testing organization are indispensable instruments for a CIO. They provide the necessary support to manage the complexities of IT projects and ensure the delivery of high-quality products. By leveraging the capabilities of a PMO and a Software Testing organization, a CIO can effectively thrive.

THE COLLABORATION OF QA, BA, PM, AND ARCHITECTS IN IT PROJECTS

In the realm of Information Technology (IT), the collaboration between Quality Assurance (QA) analysts, Business Analysts (BA), Project Managers (PM), and Architects is crucial for the successful delivery of a project. Each role brings unique skills and perspectives to the table, and their effective collaboration can significantly enhance the quality and efficiency of IT projects.

The Roles and Their Interactions

Quality Assurance (QA)

QA analysts are responsible for ensuring that the product meets the defined quality standards. They work closely with all other roles, providing feedback and identifying potential issues that could impact the quality of the product.

Business Analysts (BA)

BAs is the bridge between business needs and technical solutions. They work closely with the PMs to define the project scope and with the Architects to ensure that the proposed solution meets the business requirements. BAs also collaborate with the QA team to define the acceptance criteria for the project.

Project Managers (PM)

PMs are responsible for planning, executing, and closing projects. They work closely with BAs to define the project scope, with Architects to plan the project timeline, and with the QA team to ensure that the project meets the quality standards. PMs also play a crucial role in facilitating communication between all roles.

Architects

Architects are responsible for designing the technical solution for the project. They work closely with BAs to understand the business requirements, with PMs to plan the project timeline, and with the QA team to address any technical issues that could impact the quality of the product.

The Synergy

The synergy between QA, BA, PM, and Architects is what drives the successful delivery of IT projects. Each role complements the others, and their collaboration ensures that the project is well-defined, well-planned, and well-executed.

Planning Phase

During the planning phase, the PM and BA work together to define the project scope and objectives. The Architect then designs the technical solution, and the QA team defines the quality standards and acceptance criteria for the project.

Execution Phase

During the execution phase, the PM oversees the project progress, the BA ensures that the solution meets the business requirements, the Architect addresses any technical challenges, and the QA team continuously monitors the quality of the project.

Testing Phase

During the testing phase, the QA team conducts various tests to ensure that the product meets the defined quality standards. The BA, PM, and Architect support the QA team by addressing any issues that arise during testing.

Delivery Phase

During the delivery phase, the PM coordinates the project closure, the BA ensures that the solution meets business needs, the Architect ensures that the product is technically sound, and the QA team ensures that the product meets the quality standards.

Conclusion

In conclusion, the collaboration between QA, BA, PM, and Architects is a key factor in the successful delivery of IT projects. Each role brings unique skills and perspectives to the project, and their effective collaboration ensures that the project is well-defined, well-planned, and well-executed. This synergy leads to high-quality products that meet business needs and provide value to the stakeholders.

IMPLEMENTING A PROGRAM/PROJECT MANAGEMENT OFFICE AND SOFTWARE TESTING ORGANIZATION

The successful implementation of a Program/Project Management Office (PMO) and a Software Testing organization is a strategic move that can significantly enhance the efficiency and quality of IT projects. This section outlines the steps and considerations involved in setting up these crucial entities.

Setting Up a Project Management Office (PMO)

Define the Mandate

The first step in setting up a PMO is to define its mandate. This includes outlining its roles and responsibilities, such as overseeing program & project management standards, ensuring consistency in project delivery, and facilitating communication among stakeholders.

Establish Roles and Responsibilities

Next, establish roles and responsibilities within the PMO. This could include roles such as PMO Director, Program Manager, Project Manager, and Project Coordinator. Each role should have a clear job description and set of responsibilities.

Provide Training and Support

The PMO should provide training and support to project managers and other team members. This could include training on project management methodologies, tools, and best practices.

Setting Up a Software Testing Organization

Define Testing Methodologies

The Software Testing organization should define the testing methodologies to be used. This could include manual testing, automated testing, performance testing, and security testing.

Create Test Plans

The testing team should create detailed test plans for each project. These plans should outline the testing objectives, the types of tests to be conducted, the testing schedule, and the resources required.

Conduct Regular Testing

The testing team should conduct regular testing throughout the software development lifecycle (SDLC). This includes unit testing during the development phase, system testing after the integration of different components, and acceptance testing before the product is delivered to the client.

Fostering Collaboration Between PMO and Software Testing

Organization

Organizations can create a pleasant-sounding atmosphere that maximizes positive and impacting culture which can then strengthen both the PMO and software testing teams. I remember implementing a Quality Assurance Center of Excellence (QA CoE) for a large Fortune 100 bank within the PMO organization. I asked for the backing of the CIO before I started implementation, and he voiced his support at a company town hall with hundreds of our IT employees attending. He called out as one of his agenda points: "this is the year of Quality and please support John's implementation for the QA CoE within the PMO." This tremendously helped to launch my initiative in a positive manner with less resistance to change.

Regular Communication

Regular communication between the PMO and the Software Testing organization is crucial. This can help identify potential issues early and ensure that the project is on track to meet its objectives.

Joint Review Meetings

The PMO and the Software Testing organization should hold joint review meetings. These meetings can provide an opportunity to discuss the progress of the project, share feedback, and make necessary adjustments.

Conclusion

Implementing a PMO and a Software Testing organization requires careful planning and execution. However, once established, these entities can significantly enhance the efficiency and quality of IT projects. They foster a culture of continuous improvement, adapt to changes in the business environment, and ultimately lead to the successful delivery of high-quality IT projects.

THE IMPACT OF AI TO BOTH SOFTWARE QUALITY ASSURANCE AND PROJECT MANAGEMENT

In the field of Software Quality Assurance (SQA), the blend of AI has substantially enhanced productivity and accuracy. AI-powered tools preserve automate repetitive testing processes, allowing QA teams to focus on more complex and strategic tasks. These tools can analyze vast amounts of data to identify patterns and anomalies, ensuring a higher level of consistency and reliability in software testing. For example, AI-driven test automation can quickly detect defects and potential issues, enabling timely corrections and preventing bugs from reaching the final product. This leads to improved software quality and a reduction in the time and effort required for manual testing.

Automation within software testing has existed since the 90's and I believe that this was one of the first areas where AI and its automation efforts started to replace human software testing functions and was effective. AI's impact on QA also expands to predictive capabilities. By studying historical data and learning from past events, AI with machine learning (ML) can forecast probable defects and issues before they occur. This proactive approach allows QA teams to address problems early in the software development life cycle (SDLC), reducing the risk of costly post-release fixes. Additionally, AI can improve test coverage by creating test cases based on user actions and usage patterns. This confirms that the most essential and applicable scenarios are completely tested, further refining the overall quality of the software.

In the area of Project Management, AI has conveyed significant developments in productivity and decision-making. AI-driven project management tools can examine historical data to forecast project outcomes, identify prospective risks, and enhance resource distribution. These tools present real-time insights into project execution, enabling managers to make educated choices and modify strategies as needed. For example, AI can identify holdups and propose ideal task cycles, ensuring that projects stay on the path and are finished within the anticipated timeframe. This not only improves efficiency but also leads to more successful project results. Additionally, AI has transformed communication and teamwork in Project Management.

AI-powered tools can automate scheduled everyday jobs such as planning meetings, sending reminders, and updating project statuses, freeing up beneficial time for project managers and team members. These tools can also assist continuous communication by combining with various collaboration programs and providing real-time updates and notices. This safeguards that all stakeholders are kept updated and united, reducing the risk of miscommunication and improving over-all project management. By leveraging AI, project managers can focus on higher-level decision-making and strategic planning, ultimately driving advancement and realization.

HOW POOR MANAGEMENT ABILITIES CAN JEOPARDIZE SOFTWARE QUALITY

I have experienced several factors concerning fear and communica-tion, which can adversely affect the quality of IT projects. As a for-mer VP and Director of Quality Assurance, I have seen circumstances where the political fear among the IT teams combined with a lack of communication creates an increased risk in project quality. Perhaps communication and fear are correlated or perhaps communication is a factor which controls fear. Nevertheless, it's an issue which needs to be addressed in Information Technology projects.

If political fear and business communication issues manage proj-ects, quality will always be an issue. Machiavelli once said: "It is better to be feared than loved, if you cannot be both." In business, we are not looking to be friends with people or "loved" especially as a Qual-ity Analyst. We are working each day to do our duty. It seems that individuals who do not embrace quality assurance practices impose a lack of communication and fear on others in projects.

I have seen fear manifest in two ways: 1.) Fear towards quality in that it will overturn the "power" that a manager holds, and 2.) Man-agers display fear towards others (to safeguard their position) because they do not follow quality practices.

Fear is a way to persuade others, but it should not persuade quality. This is why the quality department often needs to be separated from the pack. Biases should never influence a QA department. Clear open-minded individuals who are devoted to the company's clients, products, processes, and wellbeing should be managing it. Often, a manager overlooks quality assurance work which in turn discourages those working in the department. The bottom line—the manager that the QA Department reports to should fully agree and understand the impact and importance of quality assurance in a company.

When it comes to NASA's space shuttle disaster, what was the true cause? Was it because of fear, communication, or both which caused the "O" ring to have a defect?

Often, it is the lack of communication that causes a problem. This is because those who are "fearful" tend to manage others based on information, but they do not communicate this information well to others. For example, I worked for a company that tends to have software emergency builds a few times a week. The lack of communication (not sharing known issues) to ensure continual product testing until the end of the day is one way to control individuals. Another example is when an individual starts working at a company where previous individuals did not share software product information. These individuals feared that if the information was shared that they could lose their jobs or someone else may gain the upper hand.

Managing by fear cuts the lack of quality practices. I often see managers too worried about meeting deadlines, client expectations, and project costs. To rectify this problem, the manager must understand that certain expectations must be met before the pressures of fear come into play. Certain software quality procedures need to be met when information is formed to assure quality. These procedures include 1.) having a central knowledge base within the company, 2.) ensuring business flow, 3.) implementing project quality assurance practices and procedures, 4.) taking time for project planning, 5.) documenting the project, 6.) having a clear understanding of the client's expectations, and 7.) having a clear understanding of project goals.

1. Central Knowledgebase

Lack of information causes issues. These issues can reduce quality. When issues or important information are known, it needs to be communicated to the "whole." Finding a way to openly manage the issue, communicate it, and document it is very important to ensure that the issue is resolved.

2. Business Flow

Expectations need to be set in the beginning. When certain actions are made the agreed upon process needs to be followed. Having a good, documented business flow helps create proper communication within the company. This works hand and hand with the quality procedures in the next step.

3. Project Quality Assurance Practices and Procedures

To ensure that the change, request, or development is dealt with in the proper manner, guidelines need to be set and met. These guidelines and procedures need to be tracked, tested, and audited to ensure they track the company baseline. Example testing is also needed to ensure that the requirements of a given software project meet the request from the client, or that the procedures are being met by all departments throughout the development of the software.

4. Project Planning

Project Planning is necessary to ensure project success. Managing the stakeholders' start and end dates, documentation, delivery, and expectations is crucial for a successful project. Project planning needs to be met as soon as possible in the project lifecycle to meet all the expected criteria.

5. Project Documentation

Project documentation is crucial for the communication of deliverables from department to department or to the client. Certain expectations are to be met. Knowing what documentation will be used and delivered is crucial information for internal IT business communication and externally with client communications. Knowing what documentation is required from the company is key. One way I've done this in the past is by creating a documentation matrix. The matrix listed the required documents based on the client, project size, or project cost of what is required for each department. This helps key out the expected deliverables.

6. Clear Understanding of the Client's Expectations

A clear understanding of the client's expectations is key to having proper communication from external to internal involvement. Client expectations should be mapped out in a requirements document. Making sure that the requirements meet the client's expectations is key. In my experience, I've found that it's best to have the client sign the final requirements document. A formal document needs to be well written to address any possible action along with its requirements. Dependencies, risks, and contiguities are important information to support the requirements. These all need to be reviewed and signed off on by the client.

7. Clear Understanding of the Project Goals

Both the company and the client have two separate objections but must conclude under the same goal. The company and the client need to ensure that their goals align with the same end result! Scope and objections can change over the project's timeline, but the goals must be clear and always communicated if any changes are made. This is why all the previous steps are important for the bottom line.

These steps have helped me increase quality assurance practices and communication. When these are mapped out, fear will no longer be a problem because it's typically present when no process exists. If something is mapped out and communicated, no problem will exist except meeting the end goal!

Quality standards need to be accepted and followed from the top down. If quality assurance is not accepted from the top down, then as a second quote from Machiavelli indicates: "It is much more secure to be feared than to be loved."

STRUGGLES OF INFORMATION MANAGERS: A QUALITY PERSPECTIVE

Information Management has seen a breadth of change in the past ten to fifteen years. Different technologies, faster networks, quicker computers, and brighter individuals have revolutionized the way we view IT. As a former VP and Director of Quality Assurance, I have noticed a consistent downward trend within companies of all sizes regarding Information Management. Issues dealing with processes, requirements, change management, quality with ownership and direction are now brushed aside with little to no importance attached to them. The root cause of such comes from the American business model—increase the bottom line as much as possible by cutting costs and expenditures. Unfortunately, the practice of maintaining quality throughout production is typically sacrificed as a means to this end.

I've found that one of the underlying reasons why an IT corporation may decide to remove (or completely overlook altogether, in some cases) the steps associated with quality production is because these steps are wrongly regarded as the most lucrative to relinquish. This kind of mentality has been a constant, problematic issue for me throughout my past employment. It starts with a lack of direction and ownership where it is most necessary. A blind eye is usually turned to the idea of hiring an individual whose main function is to take ownership of overseeing company and department processes, procedures, requirements, change management, and quality. Of course, these four aspects are imperative for the creation of a suitable product or service. When these are combined with sufficient support from a "go-to" person, the scope of the project will most likely turn out to be a success.

To their credit, IT businesses generally do a satisfactory job at filling the main positions in the corporation: managers, developers, network individuals, salespeople, and help desk/support. But these roles don't always provide sufficient support for the project and clients.

Similarly, the need for fluidity in process and procedure is almost never addressed. Processes outline the business flow from department to department, the types of internal and external documentation that are needed, and the rules of how to conduct business in every division.

When business transactions are made, all stakeholders must be aware of the timeline from start to finish, as well as follow-up procedures. Executives in the majority of companies I've worked for did not find this an important task due to the fact that it is not directly connected to a monetary payoff. Instead, they believed that changing the way the IT process was handled would be too costly due to the amount of time it would take to revamp it. What my previous employers didn't understand was that the long-term benefit of such a change greatly outweighed the time that would be lost during reorganization.

Issues that can occur because of a poor IT process include ineffective or misleading communication between departments and/or clients, lack of project responsibility and flow from start to finish, workflow documentation issues, aggravation on the part of clients and various departments when specific documented material is expected and then not received, lack of employee knowledge of supporting business tools and how they are to be used, and the list can go on and on. The one saving grace of a weak process or procedure is the kind of technology that an organization uses to support it. On the other hand, technology can create an additional problem, because before it can be used it needs to be centrally managed and agreed upon by employees of the company. If a greater importance were to be placed on the process in the first place, rebuttals such as "I did not know" and "It's not our responsibility" would be less of an issue. But as so often happens in business, corners are cut, and the ants are left scrambling to pick up the pieces.

Corporations that have a few concrete requirements for each service or product are another trend I have seen time and time again in the IT business world. Often, the creation of legitimate documents that outline requirements is not a high priority. A reason behind this is due to the fact that analysis of requirements and design takes time to generate—time that, again, is not directly tied to a profit. If a client's objectives for the project do happen to be noted, they are often written in erasable marker on whiteboards or typed in emails, but almost never put into a formalized document.

The adherence to clear-cut requirements during the analysis phase can assist in troubleshooting any issues that could occur before development. For example, if changes do arise within the development process, the client must review the requirements, decipher the best course of action, and provide proper authorization for the change. Therefore, critical errors will not be easily overlooked, and traceability is created from analysis to development, from QA to the client.

Change management is an important part of both the business and technology lifecycle. When the function or condition of a technology changes, it is up to the Change Manager to ensure a smooth transition and implementation. Without proper change procedures, too many people have their "hand in the cookie jar." Having checks and balances of how a situation can be changed and who is authorized to change it is ideal. An example of this is as follows: when a problem occurs in a production system, the developers who are in charge of creating a fix are not authorized to fully implement it. Instead, the developers must go through the change management department, which ensures that changes are made properly. This division also must document any and all modifications to the system for auditing purposes or in the rare case that the system needs to be reverted. Many tools are available to aid with change management, but it is important to acknowledge that this practice incorporates aspects of the process, documentation, software, security, and hardware.

I have spent most of my career as a quality assurance professional, building up the skills to become an expert in the field. Unfortunately, what I have found is that an independent Quality Assurance department is often ignored, and all effort is concentrated on how to cut costs instead of risk mitigation. Within the IT profession, QA can be realized in two ways: quality in business, and quality in development. Quality Assurance is truly an asset to the way a business is run as well as to the condition of a product as it becomes ready in production. The purpose of Quality Assurance in an IT organization is to certify that all of the defined business practices/processes, requirements, and change management procedures follow the correct business function. Sometimes within a development lifecycle, concerns may arise regarding the timely execution of various phases. It is the responsibility of the Quality Assurance team to find these issues before they become a problem or otherwise fix them when they transpire.

Much of QA's time is spent adjusting and auditing business processes and standards and correcting process issues that the corporation shouldn't have made to begin with. Most of the organizations I've worked for have cut corners in order to save money and expect one person to be pulled in two or more directions. I believe that the roles of the QA professional need to be separated from those of direct profit generation. Biases will exist if an individual is both a developer and quality analyst and not enough time can be aptly devoted to either role. The function of Quality Assurance is to review and analyze business requirements, create test plans, execute the test plans against the developed product, report and analyze issues that were found, and create metrics from the actions performed against what was tested. Often QA professionals are thought of as "testers" instead of what they truly are: analysts. Due to the fact that the need for a quality process occurs after the development stage, QA is not viewed as an essential division. Bringing in QA early in the lifecycle, especially during analysis, will help decrease risks associated with the project.

Presently, I have observed two extremes in relation to Information Management. One is where companies decide to invest in quality processes and procedures up-front to cut down on potential risks. Second is where an IT firm spends too much attention to "billable" and "non-billable" hours without devoting enough attention to their product or services. All too often, the latter prevails in America's IT corporate world. There must be a wake-up call for all Information Managers to see the implications of how they conduct business. Organizations must review every piece of the process, rather than focusing solely on profit. Of course, the bottom line is important. It's what investors thrive on. Nevertheless, if Information Technology managers continue to overlook the importance of the processes, requirements, change management, and quality control, corporations will continue to unnecessarily increase the risk for clients.

NEVER WASTE A CRISIS: REINVENTING "QUALITY" DURING AN ECONOMIC DOWNTURN CAN REAP BENEFITS

When a country's economy is in a dive, the quality of its products and services tends to go down with it, and it is no secret that Americas economic changes can happen and when it does changes happen big. We've heard news about American automotive manufacturers going bankrupt and top U.S. banks and insurance companies dealing subprime mortgage loans to unqualified homeowners. What is thought to be the best health system in the world is quickly being overrun by insatiability, becoming so expensive that the average American cannot afford adequate health care. Global economic survival still depends heavily on our country's prosperity and industrialization. Yet, it seems we have forgotten the role that "quality" played in America's successful past.

Information Technology (IT) is now the backbone of our economy, and it is one of our nation's biggest strengths. It is the quality assurance "czar" for nearly every product made by a U.S. company and every service offered. Why not refine this incredible asset and apply it across the board to increase the quality of our goods and services? This is where I see our country headed in the not-so-distant future. In fact, companies like Coca-Cola Enterprises have already begun the transformation. Coke's wireless cloud computing environment allows its supply chain distribution truckers to monitor shipments and enter product quantities into their company smartphone while on the job. These workers no longer have to manually write and turn in paper activity logs at the end of the day, saving precious time and streamlining the distribution process.

Our country has allowed the reputation of being "first in quality" to slip from our fingertips. This kind of change didn't happen overnight, nor will we be able to regain our title in a week's time. A culture of transformation must surface among technological leaders and organizations/companies—one that values the importance of quality, creating a foundation for the future. Only then can the notion of quality trickle downward through a company and eventually out into the public. The key is to create goals with quality in mind and to integrate them into the company mission.

Seven Guidelines for Reinventing Software Quality in an Economic Downturn:

1. *Keep all Software Development Life Cycle (SDLC) processes in place.* During a recession, there is a tendency to shift and offset the original scope of the project. Overcoming this tendency will keep

you from throwing away all of the hard work you've already put into the project.

2. *Continue to audit every software project.* It's easy to change priorities and let things fly when you're dealing with a crisis (any kind of crisis—not just a recession). Try to stay on track and review your project management and quality assurance processes as you normally would. If you must do something in a different way, be sure that the outcome will remain the same.

3. *Conduct a project "lessons learned" meeting and brainstorm ways to improve processes and quality where needed.* To continually make your products and services better for the customer, conduct regular meetings with your team and discuss lessons learned since the last meeting. If possible, invite a senior manager to the meeting so that he/she will understand why quality should be a priority in the company and the implications it can have in the grand scheme of things. The best senior managers and quality analysts have a vision for the future and knowledge of how to keep things running smoothly during times of change. It is their responsibility to make quality a goal within the company.

4. *Understand the* <u>*impact*</u> *that every employee has on the product and service being offered, the company, and the customer at large.* Layoffs and reducing employee benefits are popular ways for companies to save money while in "crisis mode." If there's no way around a layoff, be careful how many quality assurance employees are cut, as it may cause an adverse effect that travels straight to the end product or service. And it's also equally important to keep employee morale up when layoffs are on the table.

5. *Remember that improved technology will also be beneficial to your company when the economy gets better.* Many software and hardware vendors during these times are looking for sales and give heavy discounts for purchases. No manager should lose sight of increasing profits, efficiencies, and quality.

6. *Be careful how much you spend on items that* <u>*do not*</u> *have a direct correlation to quality improvement.* Research the benefits of the product and the return on investments that will come if you purchase it.

Making irrational decisions or changes without the right people will ultimately lead to devastation.

7. *Be careful how much you spend on items that have a direct correlation to quality improvement.* Don't spend money on anything and everything just because they claim to help improve quality. Quality starts with the people, then the process, then technology—in that order.

Now is an ideal time for the country to prioritize quality, strengthening the USA against future economic challenges. Companies that view economic crises as opportunities will benefit in the long run.

PLANNING INCREASES ROI FOR QA TESTING

Software test data, sometimes called a "testbed," can often take days or even weeks to create—especially if a software application has many permutations and records that need to be tested. IT organizations that recognize this tend to be successful because they invest time in planning their test procedures and test automation. However, some companies overlook the importance of setting up a viable environment in which to conduct testing. Software data may be intricately embedded within other systems that hold separate data of their own, or a single system may hold high amounts of lower-level functionality data within a multitude of data pockets. Either way, the data needs to be managed. So, for any company or organization faced with this problem, it is best to implement a procedure that can speed up the process.

Step 1: Review the data schema to find the required and un-required fields for each area of functionality that your testing will affect.

Step 2: Talk with the groups or departments who are responsible for maintaining and/or creating those functionalities. For example, a Business Analyst would be a good person to contact in a department where test data is created, as they would have a wide breadth of knowledge about what exactly is being tested.

Step 3: Find out which tests from your set of requirements are being run and which ones are not. Sometimes, aspects of the software that were not originally on the test plan may still need to be tested so that a particular transaction may be completed.

Step 4: Create a unified testbed structure which can be reused and built upon. Since testing will only increase as time goes on, it makes sense to create a data structure that can be used throughout the company. Such a testbed can then be refashioned and customized according to the needs of individual departments.

Step 5: Changes that are made to the system's structure, both within the database and the system's interface, need to be consistently communicated to the person or group that is responsible for maintaining the testbed. The reason for this is that all changes must align with the software's functionality requirements, and nothing's worse than receiving a system that has changes to its interface and architecture without knowing about it. When this occurs, more time needs to be taken to find out what in the testbed should be changed, and how the change will affect other parts of the system.

Step 6: Run multiple types of data transactions or functional tests within the testbed to ensure the quality of all permutations in the test plan. After the testing is completed, more data can be entered for each of the test plans to test against the system.

It is important to plan to ensure proper testing time and accurate data setup before the test data is pushed onto a system. If an application has a variety of functionalities crossing multiple systems, investing in a well-thought-out procedure may increase the application's ROI. Test plans and test scenarios alone do not encompass the fullness of testing. Test preparation is also needed to ensure the quality of the process. Thus, when a global testbed of data is created and maintained properly, it can be recycled and reused throughout the organization.

INTERNAL AND EXTERNAL STANDPOINTS: QA THEORY AND TECHNOLOGICAL MINDSETS

Individuals who hold technical QA positions may believe that there is "too much theory" within the industry. Conversely, an individual who operates externally from a theoretical point of view would say that QA involves too much technical phenomena. It is a paradox that these "micro" (former) and "macro" (latter) standpoints are both needed to appropriately implement Software Quality Assurance. In fact, the best QA engineers are able to balance the two and are highly experienced in both. It makes sense because the QA process embodies aspects of both technological science and theoretical analysis.

There are corporations across the world that have the tendency to believe that QA engineers should test until the software is just good enough to pass. This is a micro frame of mind: "Don't ask questions, just test." Furthermore, many managers downplay the role of QA employees, whom they feel are nothing more than "testers" who bog down the system when customers are waiting for a finished product. This can sometimes stem from business politics, or in an extreme case, when a manager doesn't want to get off their high horse, they don't want QA to have the power to tell them what they're doing wrong. If your company does not value QA enough to have it involved throughout the process, especially during testing when it's most essential. I suggest that you speak up and communicate its importance to management. After all, QA is truly the lifeblood of a product.

This is why, before involving any kind of technical implementation, a QA process should be outlined along with all scenarios of what "could go wrong" mapped out in detail with a potential solution to each. Someone who doesn't specialize in QA, or even a purely "technical QA individual" might scoff at this approach. Many times, what they want is to get the product out the door in one piece, even if it's missing a few nuts and bolts. This, of course, results in a flawed product that is unable to be fully tested or even reused.

I'll fully admit there was a time when I was that "technical QA individual" who would go straight for the quick technical fix. I learned quickly, however, that if I didn't spend enough time outlining the project, I would pay for it in the end by dishing out my evenings and weekends, answering customer complaints, and trying to fix whatever had gone wrong. At a previous job, I remember one manager talking to a QA tester at 3 p.m. on a Friday before a long holiday weekend. The manager indicated to the tester that he didn't care even if the guy had to buy a cot and sleep at work over the weekend, the software needed to be fully tested, pronto. Obviously, this kind of work environment puts a ton of strain on the tester. Therefore, if management doesn't support planning and process-mapping prior to beginning the project, they'll most likely have a bunch of sleep-deprived and disgruntled "testers" to deal with on Monday morning.

In a perfect world, QA would act as a balance, with the weight of the Business Analyst team on one side and the Development team on the other. Customers and management should bolster the scale by creating a solid foundation at the base—each department supports the other. In the real world, new QA professionals will often stand behind a manager and do whatever they say (they're the expert, right?) but later find themselves trapped in an endless cycle of "because I said so" business—or to be caught in the pressures of "you said so" by management. This kind of business politics should *never* be used to control quality assurance nor the implementation of accurate procedures. It's very important to have a happy medium.

Since QA is usually the last department to get involved in a project (although they shouldn't be), they are dependent upon the proper communication channels to get the job done correctly. After a process is put together and understood by all, the technical implementation of QA can exist. Some technical QA tasks could be: how database testing will be executed; what approaches and techniques will be used to begin testing; how automated scripts will be captured; how the data will be used to complement the testing; approaches to connecting multiple tests together; a way to test applications with a particular functionality (e.g., lookups, checkpoints, drop-downs, etc.).

It's one thing if QA professionals were meant to be mere "testers" and nothing more. But, as mentioned before, the reality of the QA profession is much broader. QA is not meant to be locked up in an information silo and test solely based on the methodology set by managers. Quality Assurance Analysts are not just conducting the test but are an integral part of how the testing is done from all angles of the environment and software development life cycle. This is why QA is often split into Process, Manual, and Technical QA groups.

The Process group outlines the mapping, standards, and procedures at both enterprise and department levels within an organization. The group ensures governance based on the guidelines that were set to ensure a quality product.

The Manual group defines and creates the test plans from the requirements. This team works heavily with the Business Analyst, ensuring that proper test plans and procedures are documented. They also execute tests manually, based on how an everyday user would execute the scripts. Testing is not just a function but also an aesthetic of what was designed and developed. It is also important to test with no technical or business biases by testing from the user's perspective.

The Technical group is even more diversified. Tasks may include automation, data testing, security, performance, and even certain functions like EDI testing and data encryption using compression on an application, etc. This group exists to take the burden off the other two. If one person were to conduct both manual testing and technical testing, it would be cumbersome to ensure that the correct methods were being used to validate the test.

So, in a software organization, QA professionals need to embody both the technical "doers" and theoretical "thinkers" to be well-rounded and successful. It is imperative for all departments to work together to make a positive impact. The pressures and politics of everyday business should not take more priority than the quality of the work.

STRONG SOFTWARE AND AN IT QUALITY ASSURANCE PROCESS INCREASES VALUE TO SDLC, ITIL

The quality assurance (QA) department needs to be involved with software and infrastructure changes to ensure their functionality, security, and performance.

The centralization of quality assurance plays a key role in any company. Because of this, many larger companies tend to have a dedicated department called the QA Center of Excellence (QACOE), which may be a group that is dedicated to ensuring quality for many departments or a department dedicated to assuring QA for a group.

What I have seen in my experience is that QA professionals are often thought of as mere "testers" who run manual or automated testing programs. The corporate world can sometimes forget that QA personnel are not only experts in testing but are also responsible for the capability, security, and performance of the entire software process.

There are instances when an infrastructure group, application/web group, or security group validates their own changes and environment without first running the modifications past QA. When this happens, a gamble is taken with the company's future productivity; instead of preparing properly and relying on a QA expert to affirm that the system is, in fact, glitch-free, trust is put in the assumption that the system will do what it's "supposed" to. An excuse some IT groups/departments use for employing this methodology is that by skipping the QA process, the product will reach the market faster. Although that may be somewhat true, it acts as a scapegoat for the laziness of the department.

Additionally, when a department owns a software product license, it also gives the department the right to manage the validation of the software tools and its processes. This creates a problem because the QA department should *always* be the last to sign off on the product before it goes live. Just because a tool is fully set up with all the bells and whistles for the team doing the change, it certainly does not mean that all the risk that accompanies it will disappear.

The quality assurance process is a way to balance the company's activities and enhance productivity while decreasing risk, all of which are based on the requirements delivered by the client. It also ensures that the steps of the process are conducted in a methodical way. So, any task that is developed, changed, modified, installed, or updated needs QA verification—including software and infrastructure.

Security testing and performance testing also need to be validated by the QA group. If a change is made, the individual who implemented the change may use the testing tool to validate it, but the changes should be run past the QA department. Too often, changes that are made within infrastructure security or performance do not involve QA.

With the Information Technology Infrastructure Library (ITIL) process being so hard pushed in industry today, it only makes sense to check any infrastructure change before a client calls customer support with a complaint, such as "the server is down; it is not responding," "I am receiving a 500 Internal server error," or "performance is very slow." From a security and load perspective, changes that are being made to the application's environment (database, servers, and network) must be tested by the company's internal third-party group, which is the QA department.

This is a diagram that describes the software development life cycle (SDLC) and ITIL process based on the demands that management and customer requirements place on QA.

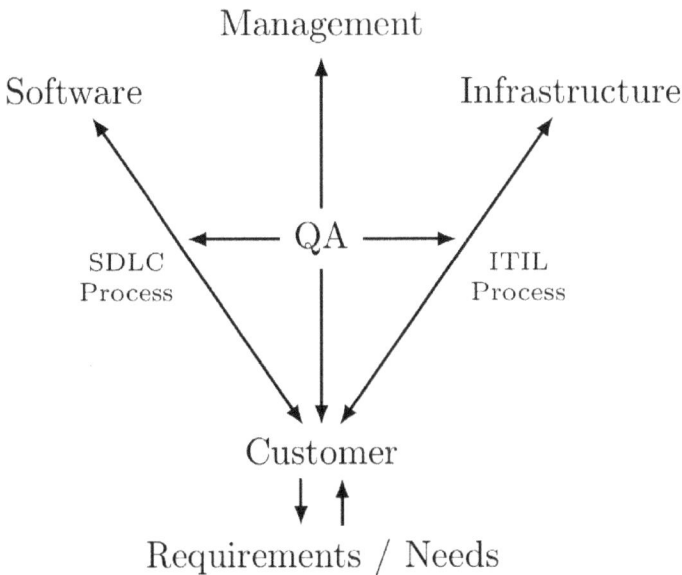

The size of the company does not necessarily indicate its maturity. Rather, maturity is measured by how a company handles the software it uses. Some corporations like to say that they are mature because they follow certain practice or use a particular tool. But—like during an interview, observation, or analysis, the result is greater than the sum of the individual parts.

I have worked in small, medium, and large companies. Many people assume that a large company would have a better process than a small or medium one simply because everything is on a grander scale, from the tools that are bought and used to the number of employees within a department. However, it has been my experience that a smaller company that has a defined process—even with a smaller group of people—can end up having a better product, service, and quality.

What's more, a smaller company generally has a faster speed to market, which helps it remain a contender in the industry. I often think of the old Roman and Greek combat methods—working together to defeat an enemy that has them outnumbered. If you stick together and have a plan, you generally end up with the sweet taste of victory. Another example is the Japanese car industry, where a quality product has been the top priority for many years. Japanese companies are based on teamwork and that teamwork continues until the job is done. *Teamwork* is also key for SDLC and ITIL. By having a centralized, non-biased QA department, quality will only increase, and risk will decrease.

Maybe certain IT work environments should be improved to accommodate a better understanding of the end goal as a collective whole. Sometimes individual departments operate with the mindset that once they're done doing what they're paid to do, there's nothing else they should worry about—when instead they should be working together on the same team to reach a common goal.

The chief information officer (software), the chief technology officer (infrastructure), the organization, and management must support the customers and ensure that the product meets their needs. QA needs to be involved with software and infrastructure changes to ensure functionality, security, and performance. Better communication will arise between these areas once they become closer to achieving a quality output together. Bridging the gap between software and infrastructure through QA will help ensure the client's success and the company's future.

Chapter Two
QUALITY PROCESS AND STRATEGY: TIPS AND STRATEGIES

"Quality means doing it right even when no one is looking."
—Henry Ford

QUALITY PROCESS FIRST. QUALITY TESTING SECOND

A problematic issue that occurs early on during a project may have a ripple effect on how "Quality" is viewed within the company, as well as how the Quality Assurance department is structured.

Often, the struggles that arise in Software Development take place during the set-up of a Quality Assurance department. Three questions come into play:

1. To whom should the QA Department report?

2. How much should QA be responsible for in terms of the systems to be tested?

3. Should quality processes and practices be covered by the QA team?

To whom should the QA Department report?

In response to this question, I will mention that QA is usually positioned under Development. If the QA department does its job to the absolute best of its ability, it must have independence from other divisions, to be separated from any biases. For example, how a project is finalized is usually considered riskier to a Developer than to a Quality Analyst. Conversely, what is considered a high-risk issue for QA might not weigh the same for Development.

My objection to Development's involvement in QA's affairs hinges on the following truth: when a deadline is looming and the project will not be fully completed on time, it's the responsibility of QA— not Development—to indicate what the risks would be if the project were to go live prematurely and whether or not to postpone its release. However, due to a "customer-is-always-first" mentality, Developers have a hard time understanding how serious the risks may be if a project were to go live before completion. For instance, a Senior Manager for Development might not think it is necessary to test a change in a production system "test environment" and so gives a minimal amount of time for QA to finalize the system's quality. When an issue does occur, it is not ultimately regarded as Development's problem but rather a Change Management, Configure Management, Quality Assurance, or Support problem.

Of course, an exchange of ideas between Development and QA could exist if power were given to a small group of individuals who are focused on a handful of given tasks within rapid development. If the QA timeline held the same weight as Development's timeline, both could reside under the same umbrella. This would allow QA to reign over the system and do its job without any discrepancies. But as is too often the case, QA is the "little guy" when it comes to risk/ quality decisions that impact the needs of clients, the system quality, the performance, and the cost-effectiveness of a project down the road.

Sometimes, QA is positioned under the same management structure as Development so that problems can be shoved in a dark corner before they become widely communicated as a "no-no" throughout the company. However, my experience has shown that filtering and blocking such information in a secretive manner will only create more problems later in the life cycle. If Quality Assurance must be positioned under a department, it should be under Operations, which has a reputation for valuing the company's well-being before anything else.

How much should QA be responsible for in terms of the systems to be tested?

How much testing coverage is required depends on the importance of the software to the consumer. Of course, one hundred percent coverage is always desirable, but how the test plans will be covered is based on severity and priority. For example, a surgical software system would need close to one hundred percent coverage because if it were flawed, it would risk human life. With bank software that deals with internal billing, however, it would be more reasonable to cut out some coverage because it wouldn't have such dire consequences. This situation must always be reviewed by middle to upper management because it can hold a huge risk to customer and end-user satisfaction. It is also another reason why I feel that QA needs to be an independent department outside of development, as stated before. The translation of risk could have several meanings, depending on the department.

Should quality processes and practices be covered by the QA team?

The third question about "process" usually doesn't present itself as an issue at the beginning of a project. Nonetheless, the process should be taken into consideration early on, or there will be a business-flow problem as the project matures. Who will manage or create the business practices and ensure that people meet engineering expectations? What standard practices, business flow, documentation, and policies will people adhere to? These questions about structure seem easy to answer, but all too often, companies spend their energy on generating profit instead of investing in the internal processes that are integral to achieving the project's end goal. Whether or not it is formalized, a process is always created before the Software Development Life Cycle is started.

In conjunction with the above, I would like to mention that Software Business Practice Quality is rarely taken into consideration—therefore leaving the QA Department to pick up the pieces and form a process on their own. Many software development companies prefer to have only a testing group. They do not see the need to have a group devoted to quality practice, because they cannot see any ROI for such a department. Unfortunately, this also allows managers to hold a monopoly over the company's business software methodology and shape the process according to their likes and dislikes. This is a big problem, for without best practices, formalized methodology, and company consensus, a project will ultimately fail in one way or another.

I have seen Sr. Managers shy away from Quality standards and procedures and instead attempt to implement a set of rules based on power and fear. Quality should never be forced upon a company but should rather be agreed upon by all individuals involved. Here are a few steps to ensure that your company is following the correct process and procedures:

1. A senior manager (or his/her subordinate) of every major department must attend each process meeting.

2. An individual who has experience with software quality process implementation should lead others in a uniform manner.

3. Personal and petty differences must be left outside the meeting room. Always keep an open mind.

4. Best practices, methodologies, and procedures must be used as reference support.

5. All departments must give insight and communication on the issues at hand equally! No department should steal the show.

6. Holding at least two meetings a week is a good way to keep a high motivation level (in some of my consultations I did these once a day until the problem was resolved).

7. Always designate someone to take meeting minutes and distribute them after each meeting.

8. Making a diagram of the new process and procedure flow is often the best way to display a methodology.

9. Find common ground for the process to fit all departments.

10. After the process is agreed upon, it needs to be communicated to the group and then used accordingly.

11. The location of the new process and procedure needs to be archived in a centralized location.

12. Suggestions and re-evaluation of the process should be conducted on a regular basis. The moment this ceases, issues may creep up again.

13. Governance is important to ensure that the process is being completed properly. Often, internal audits are conducted, and metrics are gathered to view a specific point within the project lifecycle.

14. A "time cushion" should be allotted from the very beginning of a project that will go through any new PM or QA processes. I have seen good processes fail because of a short project lifecycle, thus not allowing any time for transition.

Since "Quality" is not a tangible entity (unlike the software being tested, for instance), it is often regarded as less important. Companies that do implement a constructive process and test lifecycle, however, often become very successful.

CORRECT DEFECT MANAGEMENT INCREASES THE LIKELIHOOD OF RELIABLE SOFTWARE QUALITY

Software quality defects are an inconvenience that can stop a project in its tracks—but who should be responsible for managing them? Who should rectify the errors caused by the defect? And who should document the details of the process? Every software quality analyst and software tester should be prepared to answer these questions long before beginning a project.

In industry, the department responsible for managing and documenting software defects is often the same department that is charged with implementing the solution. The owner of the defect should not simultaneously delegate the solution because it can be difficult for defect owners to look at the issue objectively. It is therefore necessary to have a streamlined process for defect management—particularly when working with large, geographically dispersed teams. Insufficient planning may lead to poor communication, thereby creating confusion as to who, specifically, is supposed to monitor the defect (i.e., who should document and resolve the issue). I believe it is easy to manage this process in small companies or departments because of direct access to the right people. But in a large enterprise organization, the right resources tend to be stretched over a multitude of platforms, and sometimes several time zones. Therefore, the resolution strategy must be centrally managed by a process and a tool.

One way to overcome the challenges of software defect tracking is to assign each department a specific role or task. Ideally, the owner of the tool should be the software testing and software quality group. Then, a triage group or process needs to be conducted to document each new defect, along with the mitigation strategy for each defect and the owner to whom the defect is assigned. It is also very important to ensure that the correct process is implemented and used by the governance department. In large organizations, governance may be an independently operated entity, whereas in small to mid-size companies, the governance role may be wrapped into the software testing and the software quality department. Both departments should be able to clearly answer the following questions without stepping on each other's toes:

1. Who is allowed to report a defect?

2. Who should the defect be assigned to?

3. How will the defect be documented, and details changed, over time?

4. How will the status of the defect be managed (i.e., New>Open>-Fixed>Closed)?

5. Will the defects be managed by a specific defect-management tool?

6. Will the defects be automatically e-mailed to the appropriate people who will manage them?

7. Will reporting be conducted to assess the project's current status and its future progress?

8. Who should work on each defect, and which defect is the most severe, thus warranting it a "high priority?"

Once someone or a group of people is assigned and a strategy and documentation method is in place, the next step to address is how the defects will be communicated among personnel across the organization. There are many ways to accomplish this simple e-mail outlining the defects as they are found, holding regularly scheduled meetings, or drawing up a detailed hardcopy report of all errors and distributing it to the team. In many cases, the tool used for defect tracking will include notification options. All communication must inform software quality analysts, software testers, business analysts, the development/architecture groups, change and configuration management, the infrastructure group, project managers, and ultimately senior management. In other words, all of the essential departments that make up the end product or service within software development must be informed.

In some organizations, the best way to reach everyone is to hold meetings on a regular basis. The "owner" of the tool (who, again, should be the manager of software quality assurance and software testing) should be the ambassador of the meeting, offer a summary of progress made so far, and make some decisions about what needs to happen next. Using a visual defect repository tool on a projector screen, the owner should be able to facilitate active discussion among the group, thereby addressing the severity and status of high-priority tasks and managing the changes that need to be made. Large meetings may not work as well at some companies as they do at others (i.e., if e-mail conversation is typically the norm instead of face-to-face chat around the water cooler, a meeting may just take up unnecessary time), so the employee culture should be taken into consideration when deciding how to communicate defects.

If it is decided that the changes are especially severe and will require a lot of work, the Change Control Board (CCB) must be notified. The CCB is a group of representatives from nearly every department in the company. Some businesses create a CCB after a formalized process for software defect management is put into place. After the CCB accepts the changes, the decision is escalated to senior management for final approval and then carried out by the CCB through the Software Quality Assurance department.

In closing, a well–documented, communicated, and agreed–upon defect management process is essential to have in place *before* a software project has begun. The software quality assurance team should oversee the project as a gentle guiding hand, all the while keeping governance and other departments that are part of the software development life cycle (SDLC) as informed as possible. Both the software quality analyst and software testing roles are often thought of as nothing more than "bug finders." But the truth is, they maintain the integrity of the vital software products and services, and so many individuals and even entire organizations depend on them to function.

THE CONFUSION OF "QA": SOFTWARE QUALITY ANALYST VERSES SOFTWARE QUALITY TESTERS

Software quality assurance analysts and software quality testers both play an integral role in the software testing process. Software quality analysts must concentrate on delivering consistent quality across the software development life cycle (SDLC) of the company while managing procedures, processes, standards, and policies whereas a software quality tester is responsible for ensuring the overall quality of the software products.

In a perfect world, the requirements would always be clearly defined prior to any software quality testing or software quality assurance-related activity. But software quality assurance analysts and software quality testers are often given minimal direction by Project Management, Software Developers, and Business Analysts and little time within which to complete the project. The reason for this has partly to do with the organizational structure of many software technology companies.

Software Quality *Assurance* and Software Quality *Testing* are two distinctly separate functions that should be able to work together at some level but should not be rolled into a single entity. At some companies, software quality assurance and software testing are both "couched" under the same or another department—usually Project Management, Development, or the Business Analyst group. Often, the two are known simply as "QA." When this is the case, the software quality analysts and software quality testers may be assigned multiple projects that lie outside of the scope of their job and end up supporting the rest of the department that they are under instead of concentrating on the software quality process or software quality testing activity. Other organizations may not have software quality assurance analysts at all, thereby forcing software quality testers to have the "unofficial duty" of outlining process methodologies for the Software Development Life Cycle (SDLC), project management workflows, or software templates. The software quality tester's resources are stretched because they are expected to conduct testing while also doing the work of a software quality assurance analyst.

In any pure software development environment "agile," "waterfall," or even a combination of the two "hybrid" software quality assurance analysts and software quality testers must first support the needs of the business, customer, and software system before anything else. Project Managers and Business Analysts usually don't understand this because they are too concerned with the "big picture" in their departments: the overall scope of the project, its timeframe, and its cost. And although there is a direct correlation to the outcome of these factors based on the success of the software quality assurance and software quality testing process, "quality" is often left out of the equation at the back end of the project.

Since software quality assurance and software quality testing form the backbone of software development and its execution, software quality analysts and software quality testers should be free from the reins of any other department, thus forming their own distinct groups. The term, "silo effect," usually has a negative connotation when describing business practices, but software quality assurance and software quality testing are two areas where the "silo" may be the most effective way to work. In fact, IT quality is at its best when the individuals responsible for it are removed from the influence of other departments. Being separated from other departments eliminates the risk of having project managers or business analysts change the software testing process and department to suit their own needs (cost, time, etc.). It also lessens the software testers' chance of conducting testing that is biased toward any one department and reduces the number of distractions that could hinder communication between the software quality assurance analysts and software testers.

To create an outstanding product, these two entities must be able to work together while still maintaining a certain amount of exclusivity. Establishing distinct roles for software quality analysts and software quality testers allows both to work together to meet the needs of the business, systems, and customers without overlapping job responsibilities. Software quality analysts must concentrate on delivering consistent quality across the company, managing procedures, processes, standards, and policies whereas software quality testers are responsible for ensuring the overall quality of the software products in terms of their functionality, security, infrastructure, performance, etc.

In addition, a governance group should be put in place at the very beginning to monitor how the software quality analysts are delegating and performing enterprise-wide quality assurance activities, which can go a long way in ensuring that software quality activities are compliant with internal regulations. If compliance issues occur during the start of the process, they will only snowball as the software development lifecycle continues. Governance ensures that software quality assurance activities remain fluid throughout the SDLC including Project Management, Business Analysts, Change Management, and Software Quality Testing and maintains the effectiveness of each group to increase quality, production, business, and customer satisfaction. Software quality can be considered successful if software quality assurance and software quality testing are completed without any bias infiltrating the process.

The following points depict how software quality assurance and software quality testing activities should be manifested to achieve this goal:

1. Start with Process

✦ Create an SDLC that the whole company is expected to follow.

✦ Decide how the software quality assurance and software quality testing activities will be managed.

2. Group Acceptance and Agreement

✦ Every key member of the SDLC should agree, must participate, and needs to acknowledge the process that was set in step one.

3. Formalize Documentation

✦ Dedicate a specific location where all documentation will be stored.

✦ Communicate this location and the process of creating documentation to all who will need to access/perform it.

4. Establish Metrics

✦ Software quality assurance analysts should be in charge of creating audits and reports.

✦ Software quality analysts should be responsible *only* for testing reports and projections.

5. Review and Re-Establish Process

✦ Review all software quality assurance and testing activities and put together a list of "lessons learned" to be reviewed against the SDLC.

By following these five steps, the responsibilities of software quality assurance analysts and software quality testers will remain linear, formalized, and easily managed. Also, remember that although "silos" may help software quality assurance and software quality testers stay productive, the "Silo Effect" should be kept to a minimum on the enterprise-wide scale. The responsibility of "quality" is not due to any one department but the sum of the whole company. Each department must make it a goal and a priority for Quality to become successful.

CHANGE AND CONFIGURATION MANAGEMENT (CCM): SEVEN STEPS FOR A QUALITY PROGRAM

As an imperative part of the quality assurance process, Change and Configuration Management (CCM) manages and maintains all alterations to software documentation, infrastructure environments, and application codes to ensure that no unauthorized changes are made. CMM is a "portal" that all software development changes must pass through and may be run either by a group of people working together in a traditional departmental setting or in a silo where only one or two individuals are responsible for it. But, since CMM's QA role is so defined and many companies have a hard time finding a single person with enough niche experience to fill it successfully, it is usually managed by a group.

The CMM execution process must be well-defined and take the entire network, database, and software architecture into account. Here are seven key points to keep in mind that can help CMM get off to a good start.

1. ***Communication is vital.*** The group or department responsible for executing the CMM process is the epicenter of the project and will be relied upon to convey information about its progress to other groups. Knowing who, what, when, why, and how is a critical and everyday task of the CCM team.

2. ***CCM Process and Methods:***

    a. *Internal processes* for any deployment scripts, changes to infrastructure, and documentation are key. The team must be in sync so that the proper changes are made correctly, on time, and in the right place.

    b. *The technique(s) used during the CMM process* will affect changes to deployment, software environments, and documentation thereby affecting the project's overall success.

    c. ᵥ*External processes* for ensuring that any changes made by someone outside of the CMM department should always be documented and authorized by someone in the CMM department.

3. ***Having an emergency/backup plan*** is a necessity just in case an attribute is missing, or the software is not fully operable in time for the go-live deadline. The CMM team is akin to a "firefighter" in

situations such as these because they are only called to the scene when the building is burning, and someone needs help.

4. It is very important to **document all internal and external proce-dures**. All technical changes should be recorded in a single data-base as they occur so that a timeline can be created and historical information is readily available for review, if or when a defect occurs. However, the CMM team must be careful not to become the only team in charge of documentation. This duty should be shared with others so as to avoid infiltrating the recorded text with biased data—in other words, the CMM team cannot be the only group in charge of making the change and also the only group in charge of recording it in the database.

5. Make **project planning, requirements review, and analysis** a part of the CMM group's responsibility. Traditionally, CCM is only contacted when changes to a piece of software are needed, and sometimes the same changes need to be made to that piece of software repeatedly. Involving the CCM team in project planning, requirements review, and analysis from the beginning of the soft-ware development process will help them identify areas that may be prone to problems, thereby allowing the team to anticipate and prepare for the change before the issue becomes unmanageable.

6. **Change and time management** should not interfere with the CMM group's real responsibilities. For instance, any external change ac-tivity related to deployment should be conducted off-hours and on a consistently scheduled timeframe. If too many changes are made on the fly, the door opens to "requirements-creep" and a whole host of other QA issues. Internal changes should occur at daily or weekly intervals so that other departments can predict where and when they will occur.

7. **Establishing a solid reporting and control method** can help the CCM group ensure that every change is current, and all environ-ments are up to par. The method should make it simple for CMM to track and communicate present and future changes, issues that are or will be fixed, and defects that were found. For a CCM de-partment to achieve success, it must practice awareness.

Change and Configuration Management must have strong architecture and backbone process within the company for successful quality results. These seven steps will help ensure that a successful Quality Change and Configuration Management process exists within your company and individuals' projects.

HOW QA MANAGERS AND TESTERS PREP AND TEST APPLICATIONS FOR A CLOUD DEVELOPMENT ENVIRONMENT

Cloud computing is a new software service solution that holds an entire infrastructure and environment in one location, which is accessible to specified individuals via the Internet. Both hardware and software may be housed in a cloud environment. Predecessors to cloud computing include Software as a Service (SaaS), a service solution which is like cloud computing that allows users to access the software/hardware onsite or via an Internet connection; Service Oriented Architecture (SOA), which could also be accessed either onsite or offsite; and Application Service Providers (ASP), which hosts applications.

Companies that use the cloud do so to reduce expenditure while maintaining or increasing the Quality of Service (QoS). The items discussed in this article may act as a guide for professionals who are considering adding a cloud computing system to their corporate structure, thereby helping them achieve optimum performance from the service solution.

Since cloud computing systems hold vast amounts of corporate information and can only be accessed through the Internet, the availability of the Internet and the reliability of its connection are two facets of cloud computing that users depend on the most. And, although the software and/or hardware infrastructure is harbored "safely" in a cloud computing environment, the network will not always be trustworthy. There are several things that can go wrong, including a weak telecommunications signal or if your Internet provider goes out of business. For these reasons, it is a good idea to prepare a backup plan for accessing the Internet. Think about purchasing a secondary connection, in addition to your primary one, that can be available when you need it most. You may even be able to configure the system to cutover or re-route to the secondary line if the primary line experiences too much traffic at one time. If you choose to do this, just be sure that your backup wire is not from the same provider as your primary connection. Also, keep in mind that computers used in a cloud environment will never be one hundred percent reliable—there will always be viruses and other glitches that may slow down the system. So, even with the best Internet connection, a robust computer that can handle high process and memory speeds and has a large hard drive space is a must.

Cloud environments cannot be externally controlled by the companies using them, which means that problems regarding quality are difficult to fix once the cloud is fully integrated with the corporate system. In an ideal situation, a specific environment would be created for testing each application (development, configuration management, training, etc.), thereby helping to ease change, versioning, and release management along with identifying quality assurance issues. Talk with your vendor to find out if replicated environments are offered as a part of the service. In most cases, though, such a luxury typically defeats the purpose of using the cloud in the first place, as it is not very cost-effective. It is in your best interest to work with the cloud service provider to construct a quality assurance "staging" area where all of the testing, configurations, and setup are finalized *prior* to going live. You should test during this beginning phase until you're satisfied that you have the best cloud service solution to suit your needs.

When possible, test the cloud applications in a very similar (or the same) environment to the one in which it will be accessed when it goes live. The testing should scrutinize the application's performance, reliability, speed, security, and functionality. Recently, the traditional "functional testing," also known as "regression testing," is being used more than any other kind to validate the cloud environment—a one-pronged approach. To truly ensure an operable cloud environment, performance and reliability tests should take the front seat, during which probes can be used to capture statistical data and report on the consistency of the application.

The strength of the cloud's security barrier for user–protection and corporate compliance is crucial especially if your company plans to store sensitive information in the system. Formal security testing tools and even hacking techniques are some of the most effective methods for testing the security of a cloud environment. And a Disaster Recovery test will help you confirm that the vendor is reliable and responsible when faced with an emergency.

A cloud computing test plan should also be created at this point along with a detailed log of every single testing activity and issue that arises. Start by preparing a list of metrics that you want to be reported to the vendor defects or errors found, the speed of service, and its reliability. These indicators will help top management and staff address the maturity and consistency of the vendor's processes. If this is still not enough to put you at ease, create an audit of the system's log file and reports and prepare an internal/external communications plan with the vendor. Get in the habit of holding at least one weekly meeting during which you discuss future changes, status, metrics, and outstanding action items. The more communication you have with the vendor, the better they will understand your priorities, and the better service you will receive.

Once you have decided upon a cloud service solution, it is up to the QA managers and testers to make certain that the system's users have what they need to do the task at hand and continue to test applications that are needed for future projects. An area where they may need to intervene is if employees continue to save important files on their computer desktops and various locations on the corporate network instead of using the cloud system as the central source. Retrieving data can be a nightmare if you don't have access to a computer that has important information on it. Even worse is having to search through lines of folders on a corporate network for a document that someone else authored and saved somewhere. To prevent this from happening, QA managers may need to mandate that all employees use the cloud architecture when saving files. Enforcing this behavior by adjusting the read and write privileges on each computer that has access to the cloud system or by using local proprietary lockdown software might also be great solutions. For instance, you may allow data-entry personnel to use the Internet for cloud applications but prohibit them from obtaining write-access to their desktop so that data cannot be transferred to that location. Or cloud users may be prohibited from accessing public Internet sites, such as e-mail, which could be used to share sensitive corporate information. Another solution may be to use Internet machines with comprehensive capabilities instead of a traditional desktop computer, which could potentially save money and reduce the risk of data misuse.

The cloud is a great service solution to use with software products that are already stable and well-tested. It is also useful for a solution with products that will be developed within a cloud environment as long as R&D remains consistent. Be prepared to relinquish a certain amount of control of the system either way. Regardless of the reason you choose to use the cloud, however, using the techniques discussed above may help you get the most out of the service while remaining quality and cost-effective.

PLANNING FOR CLOUD SERVICES: IS QUALITY "UP IN THE AIR" WITH CLOUD COMPUTING?

Are you curious about cloud computing and wonder what it actually is? Did you prepare a list of questions to discuss with a cloud provider? Do you have an initial plan for its rollout and implementation? This article discusses the potential gaps in quality with cloud computing services, from a software quality analyst's perspective. The information provided by the **Five Steps** may assist individuals who are on the fence about switching to a cloud service and want to determine if it's a logical fit for their business.

When I first heard of *cloud computing*, I immediately conjured up the image of a person with a mass of empty confusion ballooning over his head, like a cartoon in the Sunday paper.

Turns out, I wasn't very far off. When I reviewed the definition of "cloud computing", it indicated that "cloud computing is a computing paradigm in which tasks are assigned to a combination of connections, software and services accessed over a network. This network of servers and connections is collectively known as 'the cloud.'" Now looking at the word "paradigm", "A paradigm is a pattern or an example of something. The word also connotes the ideas of a mental picture and pattern of thought." So, in a way, my original thought of a person with a mass of empty confusion ballooning over his head was accurate. Cloud computing is open to a lot of questions on how its services are being managed for good quality. It should not be subject to our own interpretation; we need to know exactly what kinds of services cloud computing offers.

The cloud method is used to categorize and manage vast amounts of data from one central location and is usually handled by a vendor or service agency that provides comprehensive management of an application, hardware, network, and security. These vendors or agencies may use functions like Software as a Service (SaaS), Service Oriented Architecture (SOA), and virtualization to perform tasks that are like those maintained by an Application Service Provider (ASP). With cloud service, the entire system is administered and supervised at an off-site environment that could be as close as a warehouse down the street or as far away as Asia or Eastern Europe.

Companies that use cloud computing do so because it allows them to add new service capabilities at any time, without the hassle of purchasing new software and training employees on the systems. But how can you be sure that cloud computing will deliver on its promise to bring your company cost-savings *and* quality? Do you really need the cloud to achieve your goals? I believe that this depends on how much risk a company is willing to take in the name of saving money. To me, anything coined as a "cloud" in the IT industry represents a network or application's unpredictability and heightened risk of error as data is transferred between two entities or locations. The key word here is *unpredictability*. From a quality assurance standpoint, this is a red flag.

Thoughts on the Cloud

The biggest problem with the cloud is that it forces the system's owner (or cloud customer) to rely wholly on the cloud service provider to manage the system, which makes it very difficult for the owner to keep tabs on the system's environment. If the cloud vendor is a third party in the system's owner network architecture (where the owner is responsible for providing service to its clients), the stakes are raised even higher.

If the network is slow, chances are that something happened on the cloud service provider's end rather than something the owner (or cloud customer) did. The issue could be related to other services that the cloud environment is providing to another client, or a controlled network demilitarized zone (DMZ). For these reasons, it is important that the entire environment for the cloud service is mapped out and a disaster recovery plan set up, just in case. A Service Level Agreement (SLA) should also be documented and enforced so as to meet the owner's needs and likewise, possibly, affect its third-party customers.

It is often harder to get out of cloud computing than it is to step into it. Sometimes, only *after* a company signs up for a cloud service does it realize its hands are tied, because all of its data and applications are then configured based on the cloud service provider's requirements. At this point, many companies find themselves feeling trapped, asking questions like, "If the system is slow or goes down completely, will customer support be ready?" and "How fast will the matter be resolved if it needs to be done ASAP?" But the real question to ask *before* purchasing the cloud service is: "How much risk to the quality of my system am I willing to accept in order to reap the cloud's benefits, and how much risk can my services and customers endure without tanking?" Also, remember that if you are not receiving what you need, the cost of these services will only increase.

Cloud computing might not be a bad solution for small to medium-sized companies, especially if the reduction in cost is significant. I would suggest that you start by using the cloud only with certain applications rather than the entire infrastructure at once. Begin small and make copious records throughout a determined time period to find out if the cost savings are worth it. With that said, I certainly would not suggest cloud computing for the entire infrastructure of larger companies, where an intricate web of network systems can be hard to manage in the first place. It's better to keep everything in-house.

Being well-informed goes a long way when considering whether to move forward with cloud computing. Do some research *before* making that decision. Of course, chances are you wouldn't even be considering switching to a new service provider if there wasn't some kind of obvious cost-benefit, but be sure to also weigh the risks and the benefits among 1.) the amount of management power you are willing to surrender, 2.) the kind of service and quality of service (QoS) that the vendor will provide, 3.) the ability to back-out if problems arise, 4.) Do they have the proper security measures in place for compliance in your HIPAA, PCI DSS, SOX, or DoD environment with proper third-party verifications and following standards from SOC 2 Type 2, NIST or ISO/IEC 27001.

Five Steps to Follow When Using the Cloud

1. **Take your time when weighing the risks of cloud computing.** Don't try to rush into making a decision, drill down to the root of the problem. What are you trying to solve by switching to the cloud? Is there another service or method that would be just as helpful/cost-saving, perhaps something that could be managed onsite? Or you may want to find companies like yours that have used/are using the cloud and ask them if they are satisfied with their service. Be sure that you have everything you need to make an informed decision, don't just look at dollar signs. Talk to more than one vendor or service agency. See what questions they ask you about your needs. Are they asking questions that are in regard to assisting with your services, or are they finding ways to get you to buy more from them?

2. **Keep a back-out strategy handy.** Make sure that you can back out of the cloud service, if needed, prior to fully implementing the system. You never know what issues may arise on your end or the provider's end, which may make more sense for you to part ways. If the company does not offer you an option to back out, it could be a warning sign of a power struggle down the road. You do not want to be locked in from the very beginning because you will then relinquish control of your system, the price structure, and your business decisions. If the cloud provider does offer a back-out plan, make sure you carefully review the documentation that they give you before you move forward.

3. **Start small by applying the cloud service only to applications that pose little risk to your overall corporate operations.** Try it out first and see how well it meets your needs. Do not risk giving up even a fraction of your infrastructure to

the vendor before setting up a demo or mock test. (But do not indicate to the service provider that you are taking them on a "test drive.") Be cautious at this stage, and watch how the vendor acts about their service; are they in the business to help their customers, or are they more concerned about making a sale? Take copious notes on how well the software performs its designated tasks, how quickly and efficiently customer support and its engineers can resolve the issue, and if they consistently communicate with you and are open and honest. Test your system around the clock to make sure that the service offerings being provided to you are top-notch and meet your goals. It is also very important for you to map out the network's architecture completely so that you always have an insider's view of the service.

4. **Ensure that you will be able to retain control over the application's environment**. If you do decide to use the cloud service, don't leave everything up to the cloud provider—you are paying them to make sure the system is being managed the way *you* need it to be! Develop appropriate documentation, measures, and metric criteria, much like you would with a mock test or demo. Be sure to keep records of all the interactions you have with the provider: who you communicate with and when along with how issues are resolved, and where exactly in the system the problem occurred. Ask the vendor how your correspondence is being tracked on their end and have a good understanding of their Customer Relationship Management process. For instance, will you be able to find an issue from the past stored in a database at any certain point in time? You might even want to go a step further and create a small internal team to keep an eye on the cloud provider. These individuals could be a part of your help desk function or from infrastructure.

5. **Maintain consistent communication with your cloud service provider.** Make it clear to them from the beginning that you wish to receive regular updates on service level status and metrics. If possible, form a relationship with someone at the vendor whom you trust. Interact with them as much as you can so that you can be sure you're both working toward the same goals. The more involved you are with your cloud provider, the higher QoS you will receive. Your involvement will also ensure that

you are not left in the dark when the vendor sees something that needs to be changed. You may want to create records of any meetings you have with your vendor to serve as testimony of what was covered, should doubt arise.

Try to fill the "cloud" above your head with facts and figures instead of question marks and exclamation points. In other words, do not be afraid to do some research before making a decision on cloud computing! Always remember: the vendor or service agency should want to be on your level, interested in your goals and objectives, and provide clear, proven solutions to help you get there.

MANUAL VS. AUTOMATING SOFTWARE TESTING TOOLS: WHEN TO USE WHICH?

Should we use an automated tool?" is a question that usually arises whenever a QA Analyst or IT firm is frustrated by the length of time it takes to finish projects and the overall cost that comes with the tools currently being used. Cost, time, and quality are not mutually exclusive you cannot be successful with one unless you are successful with the other two. Of course, the goal is always to decrease costs along with the time needed to do the job while maintaining a quality outcome.

One should first analyze how much time he or she currently invests in the creation and maintenance of automation scripts and overall speed-to-market with the current tool. Is there any way to make the tool currently being used more effective and in better alignment with the company's goals? Updating the software and re-training employees on the system to increase output are typically easy improvements that can be made. Keep in mind that just because a company has automation tools does not mean that the company is effective and efficient; instead, it's the approach that is taken with the "tools" they have to work with that determines the company's effectiveness and efficiency.

If it's decided that an automated tool does make sense to install, it should be implemented in the high-risk areas first. These are applications that tend to be the most difficult to install and maintain and are usually the high-traffic applications that see a lot of wear and tear. Applications that are used only on occasion tend to be less affected by the switch to automation and therefore should not be a priority.

When does it not make sense?

It does not make sense to use automated testing tools if, during analysis, it is found that the time needed to create, maintain, and run the scripts exceeds the time allotted to conduct quality testing of the application. If there is enough time to use and create the script, then it would be different. And even in certain circumstances, time might allow for automated software testing tools, but the resources may be a constraint. If this is the case, then a severity and priority ranking of the testing tool needs to be created for manual tests. Reviewing the rewards of cost, time, and quality is essential for the creation of manual tests.

Are any categories of automated testing tools still too bleeding-edge to adopt?

I believe that testing conducted by a software developer is too risky. The assurance of quality needs to be given to a separate group to ensure that it does not stay at the unit test level phase.

HOW TO DEVELOP A SOFTWARE TESTING TOOLS INTEGRATION STRATEGY

What do you do when your company has two or more vendors for software quality management and software test automation tools? Should these tools be integrated or remain separate?

The answers to these questions depend on several factors. For instance, you should take into consideration if your company, or single departments within your company, relies heavily on certain functional aspects of one tool and not as much on the functional aspects of another. Of course, if one tool embodies a function that you desperately need and the other one doesn't have anything that comes close, then your decision should be an easy one. But you should thoroughly review the pros and cons of each tool and be careful not to rush the process. Hasty decisions are often irrational and do not take the whole picture into account, which could cause you frustration down the road. My experience in dealing with people who are stuck in this situation seems to follow a well-worn pattern. The chosen tool often failed to meet the company's needs and/or was not compatible with other systems, which meant the end-users were unable to properly use the tools, which in turn made the tool incapable of providing the ROI it was originally intended for and increasing quality.

Just as importantly, do not allow any vendor to influence your decision. Many software vendors make promises to their clients, as any vendor would do in order to gain new business. Some guarantee an increased speed in the time-to-market and better ROI, others tout their optimized testing capabilities, and still others promise access to a wealth of quality assurance-related resources at your fingertips, like white papers and online and personal support (for an extra fee, of course). An in-house individual who has little or no contact with prospective vendors should review and analyze the tools and their processes to see which ones best fit the company's current and future needs. It may also be wise to steer clear of appointing temporary hires or consultants for this task, simply because many software vendors form partnerships with consultant agencies and their influence can sometimes trickle down the chain.

So, your ability to make a good decision without being influenced by internal or external factors is important. Nowhere is this truer than when a merger of two or more companies, or the merger of departments within a company, gives rise to the question of which software testing tool will continue to be used and which one will be dumped. The reality is that many commercial software testing tools are very similar to one another, and the decision often rests on which one is better for the environment in which it will be used. You should consider the way the tool will be used, who will use it, how it will be managed, how the tools' implementation will be communicated, and if any governance is needed to ensure its correct use throughout the enterprise.

Your decision should *not* be based on whether one of the merged departments or companies was using one tool more than another tool or if a vendor touts one feature more than another. Also, your decision should not be based solely on the requests of management (although, to a certain extent, adhering to their wishes is important). Rather, you should collaborate with the technical staff that will be using the tool daily. They have the best knowledge of what the company needs to maintain quality, and what kind of tool is likely to help the merger transition smoothly.

During your decision-making process and discussion with managers and technical staff, you may want to take the following into consideration:

1. What are all the tools that are currently being used?

2. Many tools have applications within applications—how will you integrate their functions?

3. How accessible to the users are the tools (e.g., client-server, web-based, VPN, etc.)?

4. What kinds of system applications exist in the company? In what software language, database, and framework are they orchestrated

5. How many of the tools' functions are being used? Is it worth it to even keep these tools around if you're using only one function here and there (i.e., automation scripts, data within the test management system, etc.)?

6. If your company picks one tool over another, can data be consolidated, and can automation scripts be converted?

7. Companies and their individual departments need to have a structured process for the tools' integration. If the tools are integrated without a structure, you will not be using them to the best of their ability. What are the current processes your company has in place and how can they be improved?

In today's world, software quality is only as good as the tools that ensure it. So, companies need to look at the broader scope of how those tools are put into place. Making an informed decision about the kind of tool to use and planning for its implementation will ensure that quality is practiced to its fullest.

FACTORS TO ENSURE QUALITY ASSURANCE: REQUIREMENT SIGNOFFS AND ITS TRACEABILITY

Requirement signoffs and requirement traceability are essential in the software development life cycle—especially when it comes to Quality Assurance.

In certain situations, the software development phase can start during the end of the requirements-gathering process. The problem is that developers often believe that the Quality Assurance (QA) team will always be able to find any issue with the software that may have been missed during the development phase. But, if the project's requirements were not signed off prior to moving into the QA phase, how can the QA team be expected to ensure that the product will hold up? This is a concern that can arise within basic software development life cycle techniques along with modern ones, like Agile/Extreme programming development. I have dubbed this all-too-common problem with the term, "Requirement Abuse."

There is an underlying agenda that acts as the driver of Requirement Abuse, but it shouldn't come as a surprise; companies try to reduce the time it takes to deliver a product to a client, therefore cutting corners wherever they can be cut. Adhering to requirements is usually what ends up being thrown to the wayside.

A defect is only truly considered a "defect" when it can be traced back to a requirement that shows how the problem could have been avoided if the software were developed properly. If a defect is found but is not traced back to a requirement, I consider it to be an "enhancement." I believe that, if a defect is known to pose a true problem, it would have been thwarted in the requirements for the project. A "defect," in this sense, is just based on an opinion of how much it affects the product, when it should be considered an enhancement.

When a problematic issue with a software product arises, an evaluation must be conducted to find the root cause of the problem(s). After all, if you want to prevent it from happening again, you must attack the *source*, not the *symptom*. There are several reasons that can cause a defect: incorrect/inadequate testing, a sloppy test plan and/or documentation, unclear requirements given by the Business Analyst or the customer, and overall poor development, to name a few. This is why requirement signoffs are so imperative to the vitality of a project. People often have the misconception that a project *requirement* is the same as a project *need*. Even though you can't have one without the other, they are not the same thing. Within the software development life cycle, developers are constantly adapting and measuring what the project "needs," based on the already-decided-upon requirements. Only at the end of the entire process, when development is signed off on by the client, can *needs* be considered on the same level as *requirements*.

The whole essence of QA entails that the requirements are met, and the project's needs are matched to the client's desires. When the requirements are documented at the beginning of the project, the QA team tries to anticipate any issues that may arise, according to what the client wants. This is one reason why QA should closely follow the client's mindset. Project Management, Development, and Business Analyst biases should not have a bearing on the process of meeting the requirements—but many times the QA team is controlled, in one way or another, by a timeline, politics, or a departmental change. When this happens, the project is prone to defects. Of course, defects increase the project's delivery time and, therefore, decrease customer satisfaction, ROI for all parties involved, and overall quality.

Having a customer signoff/approval is key for any and every type of software testing procedure. When it comes to the requirements of a testing process, a question that can easily arise is: "When is it [the software] good enough?" Therefore, the best way to ensure that the proper requirements (technical, functional, non-functional, and design documents) are met when the project is signed off on by the customer and approved by the business. At this point, any new changes to the requirements are considered out of scope, and a change request must then be made. If no signoff occurs, the software development process will eventually lose control of the project.

Software development firms should not confuse this approach with the kind of flawed mindset that says: "We allow our customers to make as many changes as they see fit, just as long as we can keep their business." After all, if the customer is allowed to take control of software development, then why should there even be a QA department? Sadly, many of the problems that arise within the QA testing stage are due to this. Software development companies are sometimes afraid to end what is called "scope creep"—the lengthy process of never-ending changes for fear of losing business. Clients who have been in the business for a while know that this kind of behavior is, essentially, a "Welcome" doormat. Thus, many take advantage of their software company's "eagerness to please," which ultimately results in incorrect SDLC procedures. I just recently saw a doormat that indicated "You're Not Welcomed" and I immediately thought about the aforementioned scenario. Once requirements are signed off on, a software company should not accept the customer's request for change (until the next cycle), but if they do, it should be kept at a minimum.

During many projects I've worked on, project deadlines stand firm while the client continues to ask for changes. When a software development company is caught in the rut of scope creep, the customer generally does not allow for the project's timeline to be adjusted accordingly, thereby severing the amount of time allotted for QA to conduct testing. This ultimately makes the QA Tester's job a constant hassle; what was once two or three weeks allotted to get the job done suddenly shrinks into two or three days. To make matters worse, the client still expects the software to have the full range of testing completed by the project's end. I have even heard a senior manager once say to a tester: "This needs to be tested and finalized by this weekend—even if you have to buy a cot and sleep here."

Companies that are *truly* mature and successful recognize that "client signoff first" should be a mantra that is practiced throughout the entire organization. Employees must be ready to adhere to this process even before taking on their first client. That's not to say that a preliminary development concept couldn't begin without a signoff but that the meat and potatoes of the development cycle (test design and test planning) must be postponed until the client signs off on the previous elements (namely, the requirements) of the project. This is because the development cycle is based on the requirements to ensure that both software and infrastructure activities align with each other.

After signoff, having a viable way to trace the requirements is imperative.

Creating a traceable trail of all testing activities, including the repair of software defects, ensures that the development cycle's workflow is properly documented. If a developer or business analyst has a question about the history of the project, or what was determined to be the cause of a defect, *traceability* makes certain that they can easily find the answer. Having clear documentation is also important for audits. From changes to the requirements, QA, development, and to the final signoff, the entire process needs to be proved. If not, many audits will not be deemed valid.

Before the process of Quality Assurance can begin, the following must be observed:

1.   Requirement signoff, to prevent detrimental risks from surfacing.

2. Proper documentation from both Business Analysts and Project Management, including design documents and/or technical, functional, and non-functional specifications.

Note: Be cautious of the "smoke and mirrors" approach that some PMs and BAs use to make the cycle seem to be moving along fine, but in reality, nothing is being accomplished. This goes for requirements, documentation, and milestone planning (Now, if they actually mention "smoke and mirrors," then you are in trouble!).

1. Analyze the requirements to create test plans and design a testing workflow.

2. Execute tests and report on them. At this point, QA acts as a PM to ensure the re-works are delivered on time and correctly—but do not feel that you are done. Work with development in a friendly manner and be sure that development is responsive to the issue.

3. Some time for preparation is needed (depending on your tests) to make sure that the production changes before the project is delivered to go live.

If the above issues are not addressed, risks can occur within the software development phase of the QA process. As Philip Crosby indicates, "quality is free," but at the same time, if the proper steps are not made to ensure quality *before* the testing process begins, then it will only cost both the client and the software development company more money in the long run not to mention, some disgruntled QA employees.

TECHNOLOGICAL TRENDS AND THE EFFECT IT HAS ON THE QUALITY OF MEDICAL CARE

The quality of healthcare in this country is one of the many hot-button issues debated by the 2024 presidential candidates. Traditionally, the steady decline in satisfactory medical care for Americans has been attributed to rising healthcare costs. The presidential candidates that debated the issue thrive on arguing over questions like, "Would government-subsidized health insurance improve the general quality of health of the population since more people would be willing to see their doctor when needed," or "Would it decrease the overall quality of healthcare for all Americans since demand would overpower our current supply of medical staff?"

I am not trying to get political. The purpose of this segment is not to debate the issue of health insurance; rather, I want to give you some of my insight and analysis that was conducted from data collected by an online survey. Question: Has the quality of healthcare in America increased or decreased as medical technological advancement becomes readily available? Is the healthcare industry set to steadily advance and increase in quality throughout the twenty-first century?

With an aging population, more people are needed in the expanding healthcare sector. You've probably noticed a plethora of advertising for jobs in the healthcare industry within the last five to seven years. Nursing and Physician Assistant careers are especially pushed via technical schools that offer two-year degrees. Although this is just my opinion, it seems that these schools exist for nothing more than to pump out as many graduates as they possibly can before the deficit of healthcare jobs becomes too large to adequately fill.

I'm concerned that we've adopted a flawed system—one that thrives on filling positions with people who are enamored with the "idea" of having a medical career for nothing more than the potential to make money, gain fame, and/or social status. Technical schools lure their audience with a tuition reimbursement plan that will help them pay for their education or allow them to attend virtually for free. If you had no way to go to college and someone offered you a free ride as long as you picked a medical profession, wouldn't you take the offer? Few healthcare graduates truly embrace the passion and glory of achieving internal gratification based on helping others. The result is an increasing number of medical practitioners who aren't willing to put the necessary work and time into their jobs.

"Healthcare-career-fits-all" approach is a detriment to the overall quality of medical care in this country. In order to appeal to as many people as possible, there is not as much emphasis on retaining a breadth of knowledge of medical procedure and terminology as there has been in the past. Where medical education is lacking, *medical tech-*

nology is now relied upon to accomplish the real work that was once required of a doctor or a nurse. So, if you think about it, we're not really meeting the demand for healthcare professionals at all. Instead of filling jobs with *qualified* individuals, we've simply taken the masses and thrown them into the mix to fill up space. Anyone who can press a button on a dialysis machine can basically become a licensed medical professional.

But let's take a step back.

In the 1800s and early 1900s, nurses and doctors were few. They were *specialized* and not nearly as dependent on technology as we are today. They knew the ins and outs of their profession and did their job well—even without all the current surgical technology and diagnosis equipment. Another example is the invention of the calculator. This item was life-changing because it helped people save time and energy by becoming more efficient and accurate. But at the same time, it decreased our motivation to learn and retain mathematical skills. The "information age" has made calculators a normal part of school math classes and even standardized tests. Multiplication tables must still be memorized in third grade, but what good does it do if, in fourth grade, students are allowed to use a calculator? My generation (and the generations that will come after me) is so dependent on the calculator that many people cannot even do simple computations without it.

Our dependence on technology is also prevalent when it comes to the computer systems and high-tech medical equipment used every day by doctors. Instead of a team of doctors gathering to bounce ideas off one another, there are machines that can analyze a patient's condition, perform medical tests, and dictate the diagnosis and remedy. Our ability as human beings to be independent, creative, analytical thinkers has significantly diminished because we have surrendered ourselves to automated equipment. This is and will continue to be a severe issue that we will have to deal with in the future.

When I was in high school, I worked at a small local pharmacy helping the pharmacist "push pills." Even though I had no formal medical education, I was able to appropriately inform customers of the side effects and possible interactions that their prescription may have with other drugs. How was I able to do this? The pharmacy's computer database did all the work for me. Today, pharmacists have all the information they need stored on a computer or the cloud, everything from a customer's health background to detailed information about every drug available. All they have to do is print out a sheet of paper with information compiled by the computer, hand it to the customer, and tell them to call an automated 1-800 number if they have any questions.

Later in my career, I worked in a hospital testing and quality-assuring pharmaceutical, surgical, emergency, and clinical software applications for patient information. The same question keeps popping into my head: "Will the doctors or medical experts know how to use these applications, or are the applications and robotics the experts?" The reliance on surgical robots and AI has raised concerns regarding the true abilities of doctors and even surgeons. This issue has led to moral and ethical considerations in the development of such technologies and their potential impact on individuals' lives.

An article that was published in The Wall Street Journal titled "Bedside Manner: Advocating for a Relative in the Hospital" by Melinda Beck[1] discusses the need to have relatives in the hospital to ensure the patient's care just in case of mishaps. Why should this even be necessary? The article indicated that an estimated one hundred thousand hospital patients die every year in the U.S. because of preventable errors.

The article mentioned an incident:

"A market researcher in Lake Foredt, Ill., often felt helpless while her 71-yearold father was recovering from a lung transplant in a big teaching hospital in 2005. He was faring well until he fell, hit his head, and was made to lie flat until a neurologist could evaluate him. While he waited all weekend his new lungs filled up with fluid. He developed pneumonia, then a pulmonary embolism and had three MRSA infections. He died seven months after the transplant, having never left the hospital." Having an aunt right now in the same circumstance does not give me confidence that a doctor or medical assistance would have acknowledge the right precautions

1 Melinda Beck, "Manner: Advocating for a Relative in the Hospital," The Wall Street Journal, Oct. 28, 2008.

before the infection/pneumonia occurred. The acknowledgement of the medical experts should not be done after a medical test but rather prior for preventative actions.

Don't get me wrong, I love technology! It is because of hi-tech devices such as computers that I have a job today, and I *love* my job. When it comes to the healthcare industry, science and technology have contributed to the good and the bad. Ironically, while physicians have become more dependent on technology to do their work for them, technology has had a significantly positive impact on the industry as well. There are less-invasive surgical procedures that take less time to heal and still yield amazing results. We also have advanced imaging equipment that can pinpoint malignant tumors before any symptoms are present. Nevertheless, medical professionals acknowledge that advancements in technology, developed through scientific progress and historical innovation, are integral to their ability to perform their duties effectively.

IMPACT OF THE EATPUT MODEL ON SOFTWARE DEVELOPMENT AND TESTING

The **EATPUT model**, an acronym for Event, Acquisition, Transmission, Processing, Utilization, and Transfer, is a model for analyzing information systems. It was developed by my mentor Dr. Anthony Debons from the University of Pittsburgh and has been widely used in the fields of information systems and information science. This chapter explores how the EATPUT model impacts software development and testing.

Event

In the context of software development, the **Event** phase corresponds to the identification of requirements. These requirements are the occurrences or events that the software system must handle. They can be functional requirements (what the system should do) or non-functional requirements (how the system should behave). The quality of software development heavily depends on the accurate identification and representation of these events.

Acquisition

The **Acquisition** phase in software development involves gathering the necessary data to fulfill the requirements identified in the Event phase. This could involve capturing user input, reading data from files or databases, or receiving data over a network. In software testing, this phase ensures that the system correctly acquires the necessary data under different scenarios and conditions.

Transmission

Transmission is the movement of data within the system. In software development, this involves passing data between different components or layers of the system. In software testing, the Transmission phase checks that data is correctly and efficiently passed around the system, and that no data is lost or corrupted in the process.

Processing

The **Processing** phase involves manipulating the acquired data to produce the desired output. In software development, this is where the main logic of the system is implemented. In software testing, the Processing phase verifies that the system correctly implements the required logic and produces the correct output for a given input.

Utilization

The **Utilization** phase involves using processed data to make decisions or control other parts of the system. In software development, this could involve updating the system state, displaying information to the user, or sending data to other systems. In software testing, the Utilization phase checks that the system correctly uses the processed data and that it correctly updates the system state.

Transfer

The **Transfer** phase is the action component of the system, where the knowledge generated by the other phases is implemented. In software development, this could involve writing data to a database, sending a response to a user, or triggering other systems. In software testing, the Transfer phase verifies that the system correctly implements the required actions.

In conclusion, the EATPUT model provides a useful framework for understanding and analyzing software systems, and it can also help in understanding how to manage software projects and its testing for planning, analysis, and troubleshooting. By mapping the phases of the EATPUT model to different aspects of software development and testing, we can gain valuable insights into the construction and operation of software systems. This can lead to more effective software development practices and more thorough software testing, ultimately resulting in higher-quality software.

WHEN YOU KNOW IT'S TIME TO PULL OUT THE QA PLAYBOOK: A FOOTBALL ANALOGY

Have you ever tried to explain an IT-related Quality Assurance (QA) problem to someone who has no industry background? Perhaps you found yourself in this situation while speaking to your spouse at the dinner table after work or while having a drink with a few friends who do not work in your field. If so, then you know how difficult it is for them to fully grasp the issue and how frustrating it can be to try to explain it to them.

Now, imagine the exasperation you might feel while discussing QA and the Software Development Lifecycle (SDLC) with people who work in the IT industry but have little understanding of the mechanics of their work environment.

I'll admit that over the twelve-plus-year span of my career, I've received more blank stares from IT professionals during QA business meetings than I care to mention. And the *really* sad part of it is many of the people involved in those conversations were supposedly "experts" in the field! After having this experience repeat itself multiple times, I realized an unfortunate truth: there's a host of IT professionals (even experts) who believe QA encompasses nothing more than "testing for defects," and the SDLC and its process are only "theory." The perceived ignorance of today's IT workforce is alarming, and it includes both amateurs and seasoned professionals.

If you've ever been in a situation like the one just described, you may be asking yourself: "How do I get everyone on the same page?"

Well, for all the football fans out there, you can use the game to explain how the components of a well-run SDLC and QA process should work. I used the following analogy in a real business meeting not long ago to discuss the inner workings of Software QA with a few people who were not well-versed on the subject. The explanation worked well, or I wouldn't be sharing it here. Think of it as a playbook for QA.

If IT were like the game of football, the center position on the offensive line would be a Software Developer. He is responsible for delivering the "ball" (in this case, the ball is a software product) to the Quarterback during the snap—an important action that shapes how the rest of the play will commence. Much like the quarterback, an IT Project Manager's role is to call and change plays as the game unfolds while anticipating every move the opposing team will make. The Project Manager knows that the guard and tackle positions on his team's offensive line should always be on the defensive against software flaws, which are their biggest opponent. The guard and tackle positions act as the team's QA epicenter by blocking the flaws and keeping a pocket open for the Project Manager to pass the "ball" down the field or for the running back (Associate Project Manager) to run the ball.

Of course, every great NFL team has a great group of coaches. Business Analysts are the "coaches" in IT. They create plays, or QA requirements, for the quarterback to follow and manage for the rest of the team. If a requirement is botched (perhaps the quarterback misunderstood the coach's call), a timeout should be called so that the team can huddle, brainstorm another approach to the play, and come back stronger.

Whenever a football game is won, the glory and praise usually go to the quarterback and coaches for leading the team to victory. Not much is different in the game of IT. Although a software project's success is heavily dependent upon the strength of the QA's offensive line and their ability to block the opponent, the fans (customers) typically do not give them as much credit as they do to the Project Manager and Business Analysts. For this reason, QA players sometimes do not feel a strong sense of accomplishment, even if they know the application works because of their expertise.

Some people have told me that QA is too painful, we take a lot of hits and don't get to raise the trophy above our heads for a victory lap. I think that the culture of a company's IT department, much like the culture of a football team, shapes the motivation of its members. NFL brands that are able to maintain a loyal fan base and consistent culture utilize positive reinforcement, proper training, and solid coaching to improve the performance of the team—this is especially done for the players who may be beyond the camera's view on the field but are still vitally important to the game.

When you someday find yourself in a situation that requires you to enlighten others on the practice and processes of IT, you may find this comparison of QA to football useful. The same type of analogy could be given to a defensive midfielder in soccer or a catcher and shortstop in baseball.

Chapter Three
APPLICATION LIFECYCLE MANAGEMENT (ALM): TIPS AND STRATEGIES

"To me programming is more than an important practical art. It is also a gigantic undertaking in the foundations of knowledge."
—Grace Hopper

ALM INTRODUCTION IN SOFTWARE DEVELOPMENT

Application Lifecycle Management (ALM) must be in place before any type of quality assurance (QA) activity commences. Too often, QA is conducted without any supervision to ensure that it is being done correctly. That said, the thought of moving to ALM may be daunting to some QA and project managers. In this article, we'll look at ways of establishing effective ALM and QA processes within your organization.

Outlining the effect a potential QA procedure will have on the company's revenue, sales, reputation, marketing activity, and overall quality should occur before the process is ever instated. There also needs to be appropriate oversight on the project, from which a sphere of influence can be pushed by the top corporate decision-makers down to the bottommost rungs of the organization. ALM managers frequently lack communication and guidance from executives, leading to a disorganized ALM process.

An effective way to ensure this scenario does not happen is to introduce all ALM changes and additions through Change and Control Board (CCB) meetings. C-Suite decision-makers are usually present at these meetings, so you will have their undistracted attention for at least a few minutes. It is important that they understand how a change in ALM will affect the company in the same way a technical change or software risk would. Such modifications should not be carried out behind closed doors without first consulting higher management, nor should changes be decided upon, documented, and finalized by higher management before reporting to the ALM manager. In order for any change to the ALM process to be effective, all involved parties must engage in meaningful dialogue prior to implementation. Once all parties have agreed to the changes, these can be implemented during the application lifecycle. What approaches are commonly used to introduce Application Lifecycle Management within an organization?

First, a Change and Configuration Management Team must be appointed, and an evaluation of the current situation should be conducted. Turnover of code and infrastructure environments assist with managing and directing how procedures will work. Change and Configuration Management is the centralization of all development, management, and environment changes.

Next, a Quality Assurance Team should be formed. ALM and its corresponding technology should belong to the QA Team, which ensures that the process and end product meet the standards of quality set by the organization. While ALM creates and analyzes the processes, the QA team must ensure that ALM aligns with the intended project and development in progress.

Finally, upper management must be available for communication as needed. ALM personnel, like QA analysts, should not report to a CIO or CTO. They should report directly to an individual who holds an unbiased position and should not influence the output of the project. This person is usually a comptroller or CFO whose job is to make certain that the ALM process does not adversely affect the organization's finances or the well-being of the product and customer.

Don't let a consultant or vendor talk you into believing that all companies run their ALM process the same—as this is not the case. Application Lifecycle Management processes are all unique, depending on each company's culture and procedural norms. Even companies operating within the same industry may have completely different ALM processes simply because the corporation upholds a unique mission and values statement. So, before your ALM process becomes too detailed, too comprehensive, and too confusing, take a step back and think about planning for long-term success instead of planning for quick results.

ELEVEN STEPS TO KICKOFF APPLICATION LIFECYCLE MANAGEMENT (ALM)

Application Lifecycle Management (ALM) is not to be confused with the Software Development Lifecycle (SDLC). While ALM is embedded within the SDLC, it is actually a more detailed interface that dictates how application development is conducted. In today's world, how we manage our application development can be done in several different ways. Each company, each manager, and each department has their own method of supervising what they do and how they conform with company standards when developing an application. For example, a company that develops software using a Waterfall SDLC is quite different than a company that develops software using an Agile SDLC.

Interestingly, software tools are now coming out with ALM options, but while a tool may have an ALM component, each company and vendor will often use it differently.

The methods through which a software development company uses (or doesn't use) ALM are indicative of how well their end product will meet the criteria for being successful. ALM processes should be defined and agreed upon by all internal and external stakeholders involved during and after application development. A single person or role cannot take the place of a multi-faceted process. Sometimes, projects fall into the reins of a Project Manager who dictates the process based on his/her view of what it should be—though he/she is often more concerned with the scope, time, and cost of the project. The impact of the process and ALM (how well the requirements meet the expectations of the customers) is the Quality Assurance department's responsibility. Below are eleven steps that can help you achieve successful Application Lifecycle Management.

1. **Define the roles that should be part of the ALM.** Define who is responsible for each role in information technology, software technology, quality assurance, software testing, the lines of business, and on the customer's side.

2. **Define the goals and objectives of each role.** If this step isn't followed through, then the details of each role will not be understood, and the question of "What do I do?" will never be clear. This step helps to avoid the "he said/she said" and finger-pointing.

3. **Anticipate a certain amount of internal-process dependency.** Outline and illustrate what work and which roles will depend upon your actions and deliverables. If this is not clearly documented, other people working on the project may assume that time is not important, or deliverables can be delayed or covered up if done incorrectly. A proper ALM process ensures that all levels of work are done correctly and follow quality standards.

4. **Develop a risk and contingency plan.** Every role and each step in the process has the potential to make waves if something goes wrong. So, make sure you think about how a missed deadline or objective could be rectified without a major disruption to the overall ALM process. If the risks are not understood at the beginning of the project, then you will be working in a "ticking time bomb" environment. Problems affecting time, functionality, quality, outputs, security, performance, schedule, and cost can all add to the risk of the end product, the application, and the customers. Thus, it is always better to be proactive with a constructive, reusable lifecycle.

5. **Prepare a process.** It is important that you outline a flowing pathway of communication and documentation. This is where the SDLC comes into play and defines how a company and/or department will operate.

6. **Foresee each individual's potential output.** Understand the capabilities of each person on your team and set expectations accordingly.

7. **Plan... plan again... and plan some more!** If you couldn't already tell, detailed planning is *a must* for ALM to be successful. Each role must be defined so that when a project is started, each role's objectives, dependencies, and needs are known.

8. **Know how communication affects your clients and other external stakeholders.** When you interface with a client and that client expects something in return, how will you involve them in the process? Communication between you and your customers is crucial—since the earlier your clients are involved, the more they will feel a part of the process and the more likely they will be to accept your thoughts and ideas.

9. **Set up quality assurance checkpoints to ensure your ALM process is working.** This step is usually conducted by a quality analyst, whose job is to ensure that all processes are running like clockwork (a quality analyst is not to be confused with a *software tester* whose role should have already been defined in step one). When a project is being run by a Project Manager, they often want to also manage ALM—but if this occurs, it is difficult to provide good quality assurance since there is little or no opportunity for outside evaluation. Quality checkpoints by a quality analyst help create balance within the process.

10. **Work toward continual improvement.** At the end of the first lifecycle attempt, it is QA's responsibility to delineate what went right and what went wrong in order to correct any bumps in the process. A "lessons learned" document should be prepared and presented to all of the key stakeholders, internally and externally, and discussion should ensue about how to improve the process for the next attempt. If not, then a process truly does not exist. This tends to happen if one person or team is "running the show" rather than a collective group.

11. **Provide governance.** After a constructive baseline is set, all process changes must go through a formal procedure to ensure new rules and changes within the company's adoption of the ALM

process. This is why step nine (quality assurance checkpoints) is so important. QA monitors the processes that are put into place. Governance then works hand-in-hand with regulatory in-house and external audits for standards and assurance.

A successful ALM process is one that is transparent and understood by everyone involved, including external stakeholders like customers. Every company is unique and the same can be said for their processes and technology. However, if software development organizations start using the eleven steps outlined above, they will be on their way to better ALM design and a more successful future.

IMPROVED APPLICATION LIFECYCLE MANAGEMENT (ALM) PROCESS FOR BETTER QUALITY

Application lifecycle management (ALM) is the process through which applications are managed during their lifecycle. Today, ALM usually helps software engineers create a process where there is none by using tools that could maintain an integrated process for the consumer.

ALM may be adopted in one of three ways: 1.) A product or solution provides the complete ALM process; 2.) The client creates the ALM process and customizes the product to fit their needs; 3.) A combination of ways one and two, a hybrid, where clients using out-of-the-box ALM functionality can use bits and pieces of their own process along with the vendor's product. Some ALM vendors sell their product based on the false notion that the tool will help a company achieve its goals by automatically providing efficiency, documentation, and speed-to-market. And, it doesn't hurt to have it all neatly packaged in a cute little bundle of software. The truth is, no perfect ALM solution exists—at least none that I've seen—and there are now many companies that cannot "think for themselves," thereby becoming dependent upon how the vendor provides the ALM solution. Planning, preparation, training, implementation, and governance are all keys to an ALM's success. Simply purchasing and installing an ALM tool will not enhance productivity but may end up hindering its progress.

I suggest finding an ALM implementation specialist who will help you put together the right strategy for ALM adoption in your specific situation. This will aid user-acceptance and make the most of the tool for your company's productivity. I suggest that your corporation's ALM implementation process be consistently reviewed by two different, but equally important entities. The first is Quality Management with a holistic view that encompasses the entire process from start to finish, including all individuals and departments that play any role in the creation of goods and services. Management must oversee the implementation of ALM from the very beginning since an incorrect implementation may interfere with or decrease the quality and performance of the ALM tool. If quality is lacking at any point in the design or analysis of the process, such as in the gathering of requirements for the production and/or deployment of goods and services, the QA Manager should be knowledgeable enough to recommend changes and work to correct the problem. A second review should be conducted by Quality Assurance; a more specific look at the QA department, how it works with the customer and the way in which it currently finds defects in a product. Of course, QA Analysts should also recommend improvements and work on changes if an issue occurs during software quality assurance testing. QA must be present in all areas of the business, but especially in the QA department for obvious reasons. Utilizing both of these perspectives during an organizational review is likely to enhance the integration of quality throughout ALM's implementation.

Figure 1 and Figure 2 compare and contrast the traditional ALM process to my version of an enhanced ALM process, including quality assurance and other support factors needed to fully implement ALM.

Figure 1: Traditional ALM

The traditional ALM process tends to involve project management, requirements gathering, analysis, design/architecture, development, test, and configuration and change management. These steps make up the basic ALM software development lifecycle, from the gathering of requirements to the development of an end product, service, or solution. Each phase intertwines with the next, making the whole process very easy to trace.

Traditional ALM Process

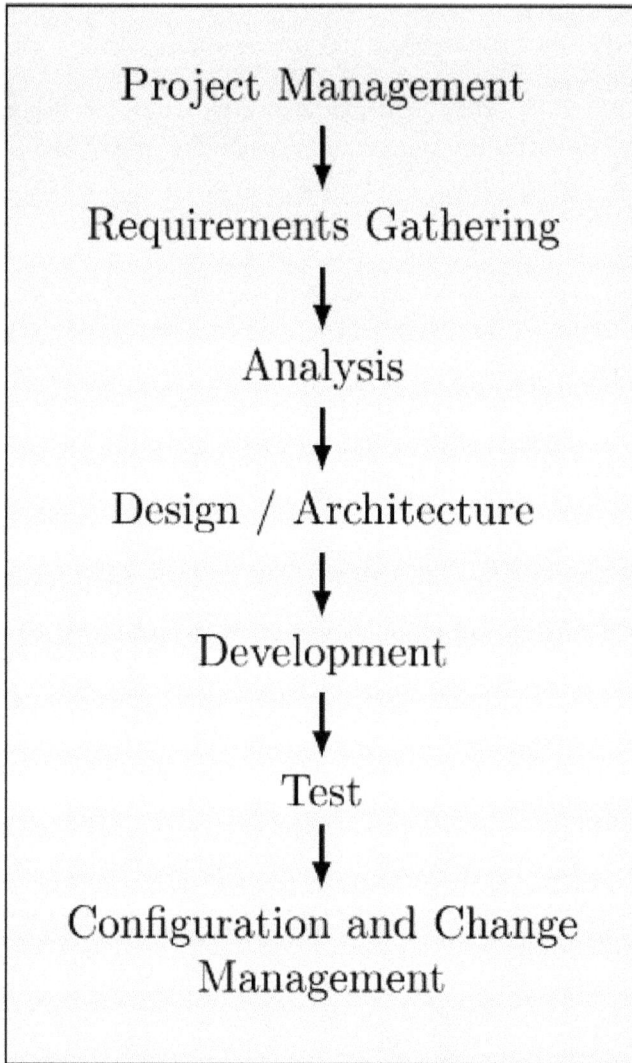

Project Management

↓

Requirements Gathering

↓

Analysis

↓

Design / Architecture

↓

Development

↓

Test

↓

Configuration and Change
Management

Although the Traditional ALM Process provides a solid base for ALM implementation, additional criteria must be interjected in order for the ALM process to become successful: Senior Management, Process, Governance, Infrastructure, Help Desk, Sales, and Quality Assurance.

Enhancend ALM Process

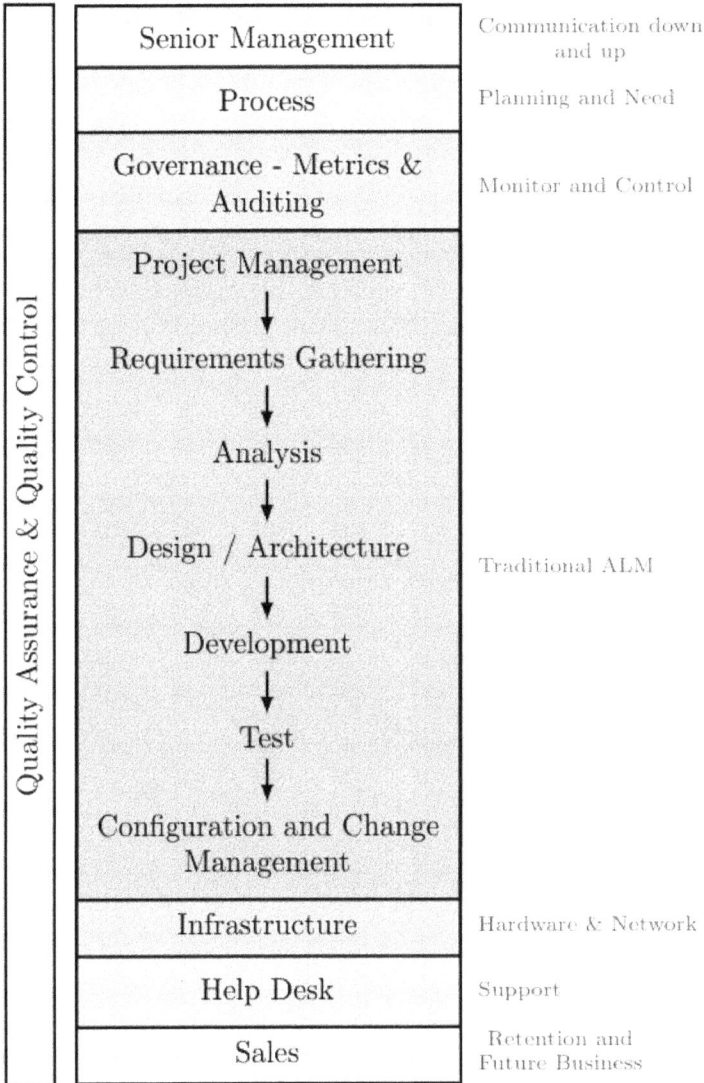

Quality Assurance & Quality Control	Senior Management	Communication down and up
	Process	Planning and Need
	Governance - Metrics & Auditing	Monitor and Control
	Project Management ↓ Requirements Gathering ↓ Analysis ↓ Design / Architecture ↓ Development ↓ Test ↓ Configuration and Change Management	Traditional ALM
	Infrastructure	Hardware & Network
	Help Desk	Support
	Sales	Retention and Future Business

Senior Management is at the top of this diagram because it is with them that any and all communication about the ALM tool begins, and they are also the enforcer of correct ALM implementation procedure. Senior management should communicate any changes in the process to their direct reports and make it clear that it is expected that the direct reports do the same with their employees. For these reasons, Senior Management must be fully committed to the ALM process and tools.

The Process Department ensures that proper planning meets the needs of Senior Management, other employees, and the company at large. Developing an implementation strategy that fits the needs of each department is an imperative goal of the Process Department since it can make or break the efficiency of the ALM implementation.

Governance ensures that the strategy of the Process Department is on track, and the expectations of Senior Management are met by all departments. The Governance team should conduct regular audits and then communicate the results and provide metrics to Senior Management and the Process Department. And, although it is in the best interest of the Process Department to make sure that all procedures are correctly followed, it is important that the Process and Governance groups remain separate entities with separate roles. The role of the Process is to create the roll-out strategy—not enforce it. If the Process team were to do both, it would be too easy for someone to make changes to the procedure and then audit the implementation's progress at the same time. To limit the risk of biased information and skewed data during an audit, it is best to have Governance deal specifically with process enforcement.

After the standard Software Development Lifecycle (SDLC) is complete, the infrastructure team should be aware of what changes need to be made and where they should occur. The Hardware and Network teams are also involved during this step of the process so they can make suggestions for change based on the influx of issues they saw during the SDLC, this is why ITIL is so crucial to the SDLC. The Help Desk is required to be informed of all changes so they can support and fix any issue that occurs during production. Its involvement with the ALM process and the Hep Desk's communication with the other levels is imperative for all who need to know if the project was a success after deployment. The key here is to document all status reports and tie them together with the traditional ALM process for SDLC.

The Sales department often gathers information on the needs of the (potential) client. Documenting and tracking this information and communicating it back up the ALM process chain, to departments that have the authority to make changes, may help increase the potential of finding a new source of revenue.

Quality Assurance and Quality Control are important contributing factors to increased revenue and ALM success that are often, and unfortunately, overlooked. The Quality Assurance department ensures that the ALM process and procedures are adequately conducted, while the Quality Control department measures and quantifies the results. Consistent reviews of each department will not only inform Senior Management of the ALM's progress, but will also assist the departments with improving where, when, and how they use the ALM tool. If all corporate departments enter the ALM process with the same goal in mind, moving as one with the same process and strategy, success will be easier to achieve.

Chapter Four

QUALITY ASSURANCE AND TESTING: TIPS AND STRATEGIES

*"Quality is free. It's not a gift, but it's free.
The 'unquality' things are what cost money."*
—Philip B. Crosby

SEVEN TIPS TOWARDS TEST AUTOMATION PLANNING

Automation." This word leads us to think of efficiency, return on investment, ways to cut expenses, and methods to bring high-quality products to market faster. The test automation process sells itself to a buyer due to the connotation of the word as just described. However, when "automation" is used regarding software testing, it needs to be taken a little differently. Implementing and applying test automation is a very strategic process and needs to be managed by an experienced testing team.

In business, where results are expected yesterday, upper management is quick to cast the blame on the software testing team when a project is behind schedule. All too often, comments like these are heard in daily status meetings: "Why isn't this software/process automated? If it were automated, testing would go much more quickly! We must automate it as soon as possible." If only it were so simple! Is test automation what is needed in situations like these? Or is it more thorough planning, less wasted time, and more knowledgeable resources that are truly needed?

I've learned that last-minute calls for automation are akin to a Hail Mary playing football. Most of the resources and time for the project have already been exhausted, and management is looking for a quick fix to get it back on schedule. By the time the project rolls around to the software testing team, it is already so far behind (usually due to glitches in development and business requirements) that no amount of automation can make up for the time lost. In addition, there is a lack of understanding within upper management about how test automation works and the preparation required to set it up. To be sure, the software test manager does not make any friends in management when he or she tells them that automation cannot be done when they want it to be!

Automation can be quite valuable if planned in advance and implemented correctly, but automating a test process will only create value for a functionality that requires minimal change. This includes the functions of a software application that have completed the development lifecycle, have been fully tested, and the results of which have been approved by the organization's subject matter experts. However, if these areas require any coding change, then automation will need to be reapplied.

As software test engineers, we must educate our managers (and, in some cases, our own peers) about what it takes to set up automation and how it works. Here is a quick list of considerations about automation that you can present to management when a project begins to ensure that everyone knows what to expect:

1. Is the development code stable?

2. Has the functionality of the application been accepted and fully signed off on?

3. Has the test automation tool been fully understood, and is it capable of being used successfully for test automation? Has a proof-of-concept POC test been done?

4. Is there a development plan in place to understand when development will be completed and when automation can begin?

5. Who will be writing the automation scripts? If the testers who execute the manual tests are the same testers who create automation tests, will this negate the productivity of utilizing these resources?

   a. Automation-driven testing > Business process optimization (BPO) from requirements to QA

   b. Requirements testing

6. Is there a maintenance plan on how after-testing is to be completed, to focus on automation areas?

7. How well does the entire project team know the software vendor and the software testing industry?

PREPARING FOR TESTING APPLICATIONS IN THE CLOUD

Doing software testing in cloud environments offers economic and scalability possibilities that are intriguing to software development companies and departments. Interest is high because purveyors of "the cloud" promise to reduce software development and testing expenditures while maintaining or increasing the Quality of Service (QoS). Do those claims hold up? How can software testers prepare for this platform shift? We'll examine that here.

You may be familiar with cloud computing, but let's just cover the basics just in case you aren't. Cloud computing is a new software service solution that holds an entire infrastructure and environment in one location, which is accessible to specified individuals via the Internet. Both hardware and software may be housed in a cloud environment. Cloud computing evolved from earlier models like Software as a Service (SaaS), which lets users access software onsite or online; Service Oriented Architecture (SOA), accessible both locally and remotely; and Application Service Providers (ASP), which host applications.

Being safe about cloud testing.

Since cloud computing systems hold vast amounts of corporate information and can only be accessed through the Internet, the availability of the Internet and the reliability of its connection are two facets of cloud computing that users depend on the most. And, although the software and/or hardware infrastructure is harbored "safely" in a cloud computing environment, the network will not always be trustworthy.

Well, it's obvious that there are a number of things that can go wrong, such as a weak telecommunications signal, a storm, or an Internet provider going out of business. For these reasons, it is a good idea to prepare a backup plan for accessing the Internet. Think about purchasing a secondary connection, in addition to your primary one, that can be available when you need it most. You may even be able to configure the system to cutover or re-route to the secondary line if the primary line experiences too much traffic at one time. If you choose to do this, just be sure that your backup wire is not from the same provider as your primary connection.

Keep in mind that computers used in a cloud environment will never be one hundred percent reliable; there will always be viruses and other glitches that may slow down the system. So, even with the best Internet connection, a robust computer that can handle high processor and memory speeds and has a large hard drive is a must.

Cloud environments cannot be externally controlled by the companies using them, which means that problems with quality are difficult to fix once the cloud is fully integrated with the corporate system.

In an ideal situation, a specific environment would be created for testing each application—development, configuration management, training, etc.—thereby helping to ease change, versioning, release management, and identify quality assurance issues. Talk with your vendor to find out if replicated environments are offered as a part of the service. In most cases, though, such a luxury typically defeats the purpose of using the cloud in the first place, as it is not very cost-effective.

It is in your best interest to work with the cloud service provider to construct a quality assurance "staging" area where all of the testing, configurations, and setup are finalized prior to going live. You should conduct testing during this beginning phase until you're satisfied that you have the best cloud service solution to suit your needs.

Best practices for cloud testing.

When possible, test the cloud applications in a very similar, or the same, environment as the one in which it will be accessed when it goes live. The testing should scrutinize the application's performance, reliability, speed, security, and functionality. Recently, traditional functional testing, also known as regression testing, is being used more than any other kind to validate the cloud, which is a one-pronged approach.

To truly ensure an operable cloud environment, performance and reliability tests should take the front seat, during which probes can be used to capture statistical data and report on the consistency of the application. The strength of cloud computing's security barrier for user protection and corporate compliance is crucial, especially if your company will store sensitive information in the system. Formal security testing tools and even hacking techniques are some of the most effective methods for testing the security of a cloud environment. Furthermore, a disaster recovery test will help you confirm that the vendor is reliable and responsible when faced with an emergency.

A cloud computing test plan should also be created at this point and should include a detailed log of every single testing activity and issue that arises. Prepare a list of metrics that you want to be reported to the vendor, such as any defects or errors found, the speed of service, its reliability, and so on. These indicators will help top management and staff address the maturity and consistency of the cloud vendor's processes.

If these steps are still not enough to put you at ease, create an audit of the system's log file and reports and prepare an internal and external communications plan with the vendor. Get in the habit of holding at least one weekly meeting during which you discuss future changes, status, metrics, and outstanding action items. The more communication you have with the vendor, the better they will understand your priorities and the better service you will receive.

In-house policing of cloud QA.

Once your company decides to use a cloud service solution, there's still plenty of work to be done by in-house QA managers and testers. For instance, make certain that the developers and/or users have what they need to do the task at hand. Naturally, there will be a need for planning the testing of applications for future projects.

Policing employees' storage habits is a key in-house project management practice. Developers, testers, managers, and users should not continue to save important files on their computer desktops and various locations on the corporate network instead of using the cloud system as the central source.

Retrieving data can be a nightmare if you don't have access to a computer that has all the important information on it. Even worse is having to search through lines of folders on a corporate network for a document that someone else authored and saved somewhere.

To prevent improper file storage, QA managers may need to mandate that all employees use the cloud architecture when saving files. They could enforce correct file saving by adjusting the read and write privileges on each computer with access to the cloud system. Another option is using local proprietary lockdown software. For instance, you may allow data-entry personnel to use the Internet for cloud applications but prohibit them from obtaining write-access to their desktop so that data cannot be transferred to that location. Alternatively, cloud users could be prohibited from accessing certain public Internet sites, such as email, which could be used to share sensitive corporate information. Another solution may be to use Internet machines with comprehensive capabilities instead of a traditional desktop computer, which could potentially save money and reduce the risk of data misuse.

In general, be prepared to relinquish a certain amount of control of the system either way. While you must be cautious about service level agreements and make sure to prepare well, cloud services can still be worth the effort. I suggest getting started with cloud services with software products that are already stable and well-tested with security certified. It is also useful to have a cloud solution that will ensure consistent R&D environment and proper SDLC with project management and testing in mind.

Regardless of why you choose to use a cloud service, the techniques discussed above may help you get the most out of the service while remaining quality and cost-effective.

WHY DO INADEQUATE SOFTWARE AND TECHNICAL REQUIREMENTS EXIST FOR SOFTWARE TESTING?

In my own professional experience, I've witnessed an increasingly common practice of Business Analysts and Project Managers providing Software Testers with very few requirements, if any, against which to test and little guidance during the creation of test plans and procedures. It is also becoming more common for requirements to be written subjectively instead of objectively, those who write the requirements are often the same people who are expected to test them. Of course, both of these situations defeat the purpose of having requirements, which is to guide and communicate the business, system, and/or customer information to Quality Assurance, Development, Infrastructure, and related departments. These days, it seems as if requirements are being generated simply to fulfill the industry expectation of having them (i.e., "we do this because it's what my boss says we're supposed to do") as opposed to fulfilling an actual need for information to communicate the system, business, or impact to the customer. And when project timelines are reduced (as they often are), so is the time available to create effective requirements. Software Testers are being forced to deliver results within a shorter time than originally planned, thereby driving them to accept whatever documentation is given to them, no matter how vague.

Why has this become such an issue lately?

One explanation may be that the high rate of turnover in today's Information Technology organizations is taking a toll on innovation and quality. Unstructured business environments can make employees more concerned with choosing colors for a flowchart rather than producing *meaningful documentation* for development, software testing, the business, and the customers. As more people leave a company, the expectations rise for employees who remain. For instance, a job description for Business Analysts used to be fairly straightforward: expertly analyze the system's impact on the business and customers, and act as an advisor to both. Now, the BA's role has shifted to encompass much more than just "analysis;" BAs are expected to spend more time in meetings and provide more technical writing, documentation, and requirements to Software Testers.

Again, it pains me to see more external documents being created within the IS/IT industry to fulfill customer requests rather than detailed, internal documents to help further the project's quality for the customer. What's really happening here is that we're putting on a façade, whether we realize it or not. As long as the customer is happy with the documents, we give them, they won't know any better—at least for a while. We are giving the customer the impression that everything is under control, while the process behind the scenes lacks quality assurance and constructive test plans and test development.

I'm not saying that making your customer happy isn't important. On the contrary, I believe good customer service and relationship-building is a crucial pillar that bolsters any IT company's success. But when "making the customer happy" comes at the expense of your products or service's quality, it is necessary to take a step back and evaluate why your BAs feel like they have to give in to the customer's every demand. Your company's reputation will eventually suffer if they don't learn how to say "no." There will always be some customers who think they know software/technology development better than your IT organization; they will argue with you to do it their way, even if "their way" is clearly not the right way. Saying "no" to these customers can be tricky, but if done in a professional manner with the right information backing up your claims (case studies from past successful rollouts are wonderful resources to share) you'll likely see the light above their heads flip on. The majority of customers do understand the value of quality done right—or they wouldn't have hired your expert organization. Furthermore, these are the same customers who won't mind if your BAs spend less time on drawing up documentation if it means that your company can deliver a better product. They are also the kind of customers who will keep coming back to your business.

I don't intend to make this entire article a lecture about finding the correct balance between customer service and the bottom line in relation to the impact of quality, so I'll move forward to my next point: What needs to happen to ensure that the business, development, software testing, and other IT roles receive adequate information to do their jobs well?

First of all, the groups responsible for following through on the requirements should not have to take on the roles of designer, analyzer, and tester all at the same time. A good deal of time must be invested in the creation, review, analysis, and signoff on the requirements to ensure that the development and software testing teams receive a legible, easy-to-understand, "quality" document. Each of these steps should be assigned to specific teams, who should spend adequate time carrying them out. But in reality, it usually all falls on the backs of software testers. They often waste a lot of time doing tasks that are not related to software testing: translating the requirements, testing to ensure that the requirements are adequate, and communicating back to the BA to ensure the quality of the requirements. It is only after these tasks are completed that software testers actually start developing test plans and test scenarios. What's worse, it seems that requirements are created behind closed doors—the people on the outside are not kept informed. The phrase "garbage in, garbage out" highlights the importance of starting correctly to achieve quality results. So, when is a process ready for software testing and development to begin?

The presentation of inadequate documentation will inevitably start a blame game, especially if quality issues are brought up by a defect or system failure. *If the system is broken, it must have been a glitch that occurred during the development phase. Or it is QA's fault because they didn't test the product thoroughly enough.* There is hardly a thought that the problem occurred because the requirements did not make sense to begin with; they were passed down through the lifecycle by people who didn't create them in the first place, and those individuals weren't given any direction by the people who *did* create them. Therefore, take care during the analysis phase to make sure you have the *right people* in the requirements meetings so that the *right informa-*

tion is passed on in the *right manner*. Pointing fingers at one particular group or department when a quality issue pops up does little but make tempers flare. Quality issues often indicate broader challenges related to organizational communication. Each department or team must uphold high standards and promptly report when support is lacking.

While communication is key, so is the ability to accept change. There is no such thing as a "perfect process" within the software development lifecycle. The customer's acceptance of the product does elevate the project's success, but one must also consider if the product was created with a quality approach throughout its lifecycle. Again, if quality assurance is valued from the beginning of a project and carried through to the end, there is a 99.9% chance of a greater quality outcome. I am not inferring that the processes and procedures should never change. No matter how "perfect" you think your process is, there will always be a need for modest change to the processes and procedures as technology evolves. In most cases, the process will get better. Organizations that resist adopting even small changes for the sake of increasing the quality of the processes and procedures risk being unable to innovate and grow competitively (the opposition usually holds to the philosophy: "If it's not broken, don't fix it.").

Car manufacturers including Lexus, Toyota, Acura, and Honda are known for their production methods. Companies such as Hyundai/ Kia, through Genesis, are working to implement comparable manufacturing practices. While Toyota faced some quality concerns a few years back, the brand has consistently supported robust quality initiatives. You may be surprised to know that many of the newest, technologically advanced vehicles out there are still being built on the same frames used ten or fifteen years ago. The frame is the "foundation" of these cars. The exterior design and bells and whistles may change and improve, and some small improvements to safety may be implemented as time goes by—but groundbreaking alterations arise only with the goal of improving quality and thus increasing sales and maintaining customer loyalty. It should not be a surprise, then, that automobile lines that constantly evolve and introduce brand new "generations" every few years typically see a decline in quality, especially in the preliminary years. In many cases, these vehicles hardly resemble the model that came before them. They lose their "foundation" and find themselves in a schizophrenic identity crisis. Much like the automobile industry, change in the software industry can be good, but it must be strategic and based on wise choices. Change for its own sake causes confusion, but purposeful change aimed at quality improvement can boost efficiency, productivity, and overall value for the organization.

A culture of quality is imperative for the creation of good requirements. Yes, meeting deadlines and staying on or under budget is important, but if those are the only things your company is worried about, then you have a long road ahead. Several aspects of quality assurance are to develop pockets of productivity and enhance communication, analysis, design, and signoff of requirements. Without proper quality assurance, the whole requirement documents will automatically become useless and virtually meaningless. To conclude, it is up to the technology offices and senior management to recognize, support, and lead requirements initiatives and other quality-related strategies. A "quality" mindset must trump a "deadline-driven" one. If not, then software projects will be more costly in terms of money, time, and, ultimately, customers. Operating from a quality mindset helps to fill in the holes among processes and procedures, ensures that the requirements are effectively working towards customer satisfaction, and improves the engineering, testing, and overall quality of the product.

NINE STEPS TO CONSIDER WHEN IMPLEMENTING SOFTWARE TEST AUTOMATION TOOLS

Software test automation tools can be very beneficial to a company's software testing organization. The aim of these tools is to increase test coverage while decreasing the cost of the actual testing function, in addition to shortening the amount of time needed to test and requiring less resources. In theory, total "automation" means no manual intervention is necessary—such actions include starting and stopping scripts, creating new scripts, and maintaining them as the project's needs change. But although "automation" sounds like a great feature for software testing tools to have, there isn't a tool on the market today that is one hundred percent, fully automated.

Software quality assurance professionals can find themselves tee-tering on a line between the amount of money they want to invest in a software test tool, and the software testing tool's capabilities. The decision often comes down to either an open-source software testing tool or a vendor-based software testing application. So, how can you be sure you're purchasing the right software test automation tool for the job?

Before you do anything, it is crucial that you put together a strat-egy prior to searching for a tool. Research and review the needs of your company and what issue(s) the tool will be expected to solve. Make a list of tasks that the tool should be able to perform, then describe how each task will improve its respective business function. Think about how large or small the tool's user base will be and keep their requirements in mind as you align the tool's functions to your company's strategy.

After you have a strategy in place, you'll be ready to choose a tool. Here are nine important factors to consider:

1. Maintenance

Longevity

Ask: How long will the vendor be able to maintain the software testing tool application? Is our company stable enough to continue maintenance if we invest sufficient time, money, and/or resources?

The application may be offered at a discounted price (or even for free), but there must be enough workforce to follow through and maintain the product.

If it's an open-source application and the open-source mainte-nance team loses their way (due to lack of resources or other prob-lems), then so might the company's ability to manage the product. But, remember, if you maintain the product in-house, then you will also need to test it in-house, which can put a strain on time and manpower, along with opening a window for bugs. Open-source products are great for environments that have long project schedules or where high-level R&D exists. They are not as feasible in environments where all that needs to be accomplished are quick, automated tasks.

Updates

Ask: Will the open-source or vendor-based application be regularly updated with new versions of the software? What if the old software has a 'bug'?

As system environments change (operating systems, plug-ins, browsers, etc.), so does the need for updates.

Updates should be conducted on a weekly—or at the very least, quarterly—basis. This is an important step to ensuring that the software testing tool's maintenance group receives excellent internal support.

2. Usability

Ask: Is the layout of the open-source or vendor-based application designed in such a way that it takes fewer actions to work on any given function? Would both a novice tester and an expert tester find it easy to use?

The layout of open-source applications can be difficult to manage and sometimes requires an in-depth knowledge of computer programming languages such as C+ or VB. And many new open-source applications are coming out with layouts that are similar to their vendor-based counterparts. Therefore, you must determine how much time the tool would take to implement and the kind of engineer it is designed for. Software testing tools have become more user-friendly as of late, but the user must still be capable of fully understanding the intricacies of the testing strategy and the application that is being tested. The tool will reach its highest potential if it is used by a "quality-minded" individual (or group of people) who understands that quality is integral to reaching the company's goals.

3. Compatibility and Integration

Ask: Does the open-source or vendor-based application have the capability to integrate with other tools, such as a defect-tracking tool? Is the tool compatible with the company's environment?

The tool's functional ability increases if it has a built-in capacity to integrate with other applications and databases. Make sure that the software test automation tool is compatible with the application being used to automate and test, the client operating system that it is running on, web browsers, and other items like change control. Additionally, the tool's compatibility is a key indicator of how accessible it will be when executing scripts. You should find out if the tool can be executed from any machine or if an automation client needs to be installed first, both of which will then determine the price of the license.

4. Manuals

Ask: Does the product include an operation manual that will help support it?

It is always a good idea to have some sort of reference material to support the tool's end-user, especially when it comes to executing a unique function specific to that tool. If reference documents are not available—as is the case with some open-source applications—you should consider creating your own. Reference guides that are compact but still indicate the full functionality of the tool are a great asset to give the end-user, especially if the original product manual is cumbersome to the everyday user.

5. Customer Support

Ask: Is adequate customer support made available by the product or vendor? What are their hours of operation? What is their response rate? Does it come at an additional cost?

Many times, what you are truly paying for when you buy a testing tool is product marketing, R&D, and support. So, keep in mind that if you are paying a lot of money for a tool, it should be top-notch—or in the end, "you will pay for what you get." Find out if the product includes customer support. Some open-source applications might claim they have customer support but read the fine print to find out if the support is only offered online or if you can actually speak to a live person on the phone. Another thing to find out is where customer support is physically located. If you have an urgent problem at 7 a.m. and you live on the East Coast, don't expect an immediate call back from the vendor's customer service center if it is located on the West Coast. Some vendors may even have offshore resources. As you test the product, it is important that you also test the customer service offerings by calling and creating service requests.

6. Online User Communities

Ask: Does the application offer an online community for users to share information with other users?

How often have you had a question that someone else knew the answer to? A product's online community is a wonderful resource to utilize when you want to find out if other users have the same problems/likes/dislikes about the product as you do. Online communities are an added value to a tool's capabilities. Not only do they help users share information with each other, but they increase brand loyalty and spur trust among current users and potential users alike. When researching if your tool has such a community, beware of the slew of unofficial collaboration communities and websites out there—for the most accurate information, go straight to the tool's official online community webpage.

7. "Center of Testing Excellence"

Ask: Who will manage the software test automation tools internally? Will there be a centralized group that will manage all department software test automation tools, or will each department manage the tool by themselves?

The method through which a tool is managed will make or break its success. Maintaining a Center of Testing Excellence makes sense—especially with software testing automation tools—because it acts as the central departmental hub where all information about the tool is stored, from management and installation to knowledge and usage. The Center of Excellence assists each department that uses the software testing tool within the departments. Sometimes, smaller groups may have independent testers who float to and from other departments. Larger groups may have permanent testers who work full-time testing in only one department. Regardless, the Center of Testing Excellence is a pillar of support—and in some circumstances, it can even act as a stand-in for departmental testers.

8. Sales

Buyer beware: the software automation testing tool you are about to purchase encompasses much more than the fluff presented by your sales representative. "Try before you buy" in order to get a feel for the technical aspects of the tool and understand how it will or won't meet your needs—these are things that your sales rep most likely never told you about. Make sure that you install the tool in a test environment, use it frequently, and then comprise a list of questions for the vendor and its technical team. Be aware of the multitude of "services" a vendor promotes; step back and analyze if it is cost-effective and if it meets the business' goals. Sales teams may use the software test automation tool as a springboard for introducing additional and often unnecessary services to their potential clients.

In some cases, it may be better to compare multiple systems at the same time and conduct your own analysis. Each tool may act differently, depending on how your company conducts testing. Don't expect the vendor or independent agencies to make a comparison of the tools for you to save time, nor should you rely on them as experts. As the end-user, *you* are an expert within your industry, and *you* need to take the time to understand if the application makes sense for *your* environment. If you speed through this process without any direction or plan, you will be stuck with a tool that meets neither the needs of the company nor the users.

As previously mentioned, it is very important to have a group of end-users involved in the decision process of whether or not to purchase the tool and any additional services. Corporate politics sometimes plays a large part in the decision-making process, and the end users are stuck with a tool that doesn't work for them and the corporate goals. So, don't be fooled by the vendor's sales team. You might end up losing in the end.

Just a note—you may want to take screenshots of the application while you are using the temporary demo, by producing a document that compares each screenshot of functionality. After the short-term license expires, you may need this documentation to help you reflect upon what was positive and negative about the tool during your final decision-making process.

9. Cost (Cost of the Tool + Indirect Costs)

"Cost" means not just the price of the tool, but also installation time, ramp-up knowledge, and other indirect costs. Some of these factors were already indicated above, like training, ramp-up speed, resources, documentation, installation, compatibility, and manuals. You should take a step back in order to account for all costs—especially the indirect ones—because sometimes the total price of the vendor-based tool may be less expensive than the open-source application and vice versa, depending on how well the open-source product was supported. To this end, indirect costs can cause a major issue and often creep up when you least expect it. For example, I have witnessed situations where applications were purchased but no budget was allocated for their installation or environment, which resulted in the tool operating in an inadequate, shared environment.

Keep these nine items in mind as a guide while you review your options to purchase software test automation tools. Make sure that you fully understand your company's goals *before* you begin the process—it will reduce the risk of choosing a tool that meets only part of the business' needs. Remember, "Quality" is the backbone of the Software Development Life Cycle. If the acquisition of a software test automation tool is rushed and conducted unsuccessfully, there's a good chance that it will cause a domino effect and infiltrate other departments. So, take your time. The better you understand what you're buying, the more confident you'll be in your final decision.

TIPS TO HELP PREP FOR UPDATES TO OPERATING SYSTEMS, BROWSERS, AND MOBILE DEVICES

The rapid evolution of the information technology industry brings with it many benefits. Smartphones, for instance, are compact, convenient, and bring immediate gratification through instant email, ring tones and alerts, text and app messaging, internet 5G access, Artificial Intelligence (AI), and now, movies and live streaming television. Of course, don't expect that your new Apple iPhone or Google Pixel will remain top-of-the-line for long. Apple or Google will have a new, juiced-up version out faster than you can text a smiley emoticon to your best friend.

With the constant evolution of mobile and computer technologies, comes an inevitable (and unavoidable) slew of upgrades to operating systems, web browsers, and application interfaces—known collectively as "clients." These days, more corporate businesses and mobile smartphone services are dependent upon web and mobile-based operating systems. Changes are inevitable with companies like Apple, Google and Microsoft. Quality Assurance analysts, who work at companies with very complex and dynamic client-dependent systems, must be armed and ready to test these changing environments in beta prior to its official launch. More often than not, though, businesses are not ready and/or willing to take the appropriate steps needed to manage such change.

When it comes to QA testing on a client, being proactive instead of reactive will ensure that the system can go live on its intended release date. Since there are a slew of different operating systems out there (Apple, Linux, Windows, Google etc.), there is not one that works exactly like the other. Companies need to utilize multiple test environments to ensure that their applications will work on upgraded devices and on diverse browser-based interfaces. Testing should be conducted against both 32-bit and 64-bit processors, against the performance of both memory and its process speed and storage capabilities.

User environments encompass a dynamic grid for test specifications, thereby requiring the tester to meet the criteria of a seemingly endless number of permutations. For this reason, investing in a solid virtual environment or a server with multiple partitions or drives that are hot-swappable may be a viable solution for some businesses. Having a test environment within easy reach helps QA analysts and QA management stay a step ahead by being able to anticipate future problems (based on what the test specifications are) and mock-test them *before* they happen. This reduces the risk of having the end-user exposed to faulty hardware or software. The key is to always conduct more testing than is required and, in more environments, than is required.

Customers want to feel safe knowing that they can choose any environment at any time and the system will work. It is often a good practice to create an environment requirements matrix with the operating systems (column) and browsers (row) listed. Plug in "X's" and color-code the matrix to reflect high and low-impact test environments.

It also makes sense to work closely with the R&D team at the parent company of the OS (i.e., Linux, Google, Microsoft, or Apple), keeping in regular contact whenever possible. Doing so will ease the process of testing the applications you intend to run on their system and help you become more knowledgeable about the changes your company needs to make to comply with the upgraded system's requirements—thereby reducing your share of risk associated with the change. Furthermore, if your company has a database with user growth and client activity statistics, you may want to carefully review it prior to testing so you have an idea of the efficacy of the current system compared to the new one. Testing not only functionality but also the performances and the security is also critical.

In the IT industry, the extra work involved with being prepared to deal with change will pay off often. Remember, the customer ultimately expects your company's technology to provide them with the proper information and accuracy needed to run a business. It is always frustrating when a system goes down—but it will hurt even more if a customer refuses to continue using your company's service because of an issue that could have been avoided by being a little more comprehensive while testing the changing environments for operating systems, web browsers, and client applications to interface with.

AN IMPERATIVE INGREDIENT IN SECURITY AND PERFORMANCE-TESTING: QUALITY ASSURANCE

In a sluggish economy, businesses are forced to change the way they respond to consumer demand. People are increasingly cautious about where they spend their money and want to get the best product they can for the cheapest price. Thus, the corporate sector is concentrating on product performance and longevity, along with beefing up the security of their internal IT systems to stop hackers from stealing sensitive product information.

Tools that are purchased for the security and performance environments are usually licensed to a specific department instead of being licensed to the company at large (even though other departments within the corporation can use it). This is a technicality that can be abused because a software license gives an entity all the information it needs to install and use the tool and the power to implement the software as it sees fit. Sometimes, the licensed department may operate without other departments that would normally be involved in the software development lifecycle (SDLC). When something like this happens, the QA department is officially out of the picture, which does not bode well for the future of the product.

Believe it or not, I've witnessed a web application group validating and testing its own product with software testing tools and without informing QA that it was doing so. I've also seen an infrastructure group hoard performance information from QA and change *the testing results* because the sole validation and verification came from within the group.

Why does this present a problem? There is no objectivity involved in the testing process because it is all conducted internally with a biased group. The involvement of QA is imperative during the testing process because only the QA department has a variety of resources needed to effectively approach different situations and evaluate them. The issue isn't so much the fact that the web or infrastructure groups conducted their own testing of security and performance; rather, it's that they were the *only* groups that conducted testing. Just because the web and infrastructure groups are the only ones using the tool does not then mean they also own the software and license, nor is it right for them to test their own definition of "quality." This increases the chance of having information withheld from other groups that need it. It's important that departments do not become "information silos" because QA is ultimately responsible for the outcome of both the process and product.

I believe that the best results come when a group tests their own product as a whole unit, and then QA tests it again to uphold the product's integrity through objectivity. Moreover, QA should use non-traditional testing approaches, such as testing "around" the product to other functionalities that may be affected by the product's implementation, just to cover all the bases. Then the verification and results can be centrally located with all the other functional and non-functional tests for a given release or project.

A good way to centralize testing information is to create a data library for the results so that every test during the SDLC is documented and accessible to everyone. It should support software requirements by including functional and non-functional test results, along with security, performance, and infrastructure implementation or installation updates. Also, the principles of quality assurance must be woven throughout the project. I like to arrange the requirements and the test plans by displaying the high-level details of each document in a test management tool or shared network location displaying the functional, non-functional, security, and performance requirements and test plans with both positive and negative approaches. Then, I'll create another folder within the test management system or shared network location for verifications, which contains all phases of my testing along with each phase's results. Whenever a defect occurs, I note the phase during which the defect took place along with which test plan was created due to that defect. By encompassing security testing and performance testing information in the same place, everything is easy to find and navigate.

CONDUCTING BUSINESS PROCESS TESTING (BPT) IN A SERVICE-ORIENTED ARCHITECTURE (SOA)

When multiple systems are in different areas and do not share the same structural backbone, they can quickly become difficult to manage. Usually, the best course of action is to implement Service-Oriented Architecture (SOA)—the business framework as both information systems and processes—which will reveal the applications that make sense to keep and which ones require integration. The ideal result from an SOA framework is a workflow that is easily managed by the business and requires little maintenance.

SOA is a software solution that brings systems together that would not traditionally be integrated for various reasons. Sometimes, two systems can create a greater output by working together than individually, where they are known to perform slowly. The goal of SOA is to improve communication channels for the business processes and the systems that the framework manages.

Business Process Testing (BPT) can become a significant add-on in terms of SOA, especially if the application's functionality is unchanging. This would require a software quality analyst to create tests that gauge the system's logical business process and weigh it against what the system was originally intended to do. On the other hand, having subject matter experts create business tests allows the "business process" test plans to be documented without much intervention from a quality analyst, who would undoubtedly feel the need to create detailed test plans for those functionalities. Where a quality analyst's test plans would be necessary for testing the details of a database, the SOA framework is selected, championing integration, security, and performance. It is also important to note that the software quality analyst should sign off on all business process tests to ensure that it not only tests the positive functionality of the application but also its negative functionality.

Utilizing subject matter experts and BPT to test an unchanging system makes sense because it cuts the time needed to have SQA (Software Quality Assurance) intervention (depending on if the system already has "good" quality). BPT procedures that are created by subject matter experts should not be used as the sole justification for a test plan. The main purpose of BPT is to test processes, and it is a perfect method for testing against back-end changes like SOA and EDI (Electronic Data Interchange). This is also applicable to a front-end user perspective, as long as the other risks to what is being changed and implemented are tested fully by software quality analysts.

Architects hold an integral role when it comes to SQA implementation. The architect is the analyst who reviews the entire system to ensure which business systems and processes need to be changed, configured, or substituted. They then put into place the SOA framework based on how the systems should be communicating. Software quality analysts must work with this individual so that they can determine which system holds the highest risk and in what way they should perform an analysis for testing. Prioritization of the system would then be made so that the SQA can work with subject matter experts for functional testing.

MONITORING AND CONTROLLING OF APPLICATIONS AND NETWORKS SHOULD HAVE QA AND TESTING

Monitoring and Controlling of Applications and Networks are part of Quality Assurance. With that in mind, the security and performance of testing and assurance of it is also part of Quality Assurance. Far too often, companies like to decentralize these parts of QA to the departments that are responsible for putting them into place. Very wrong! While it may be ideal for their own management to have control over QA when it comes to monitoring and controlling the applications and networks in addition to everything else involved, it is not ideal for Quality Assurance and management to ensure that everything is being delivered and operated effectively.

Any organizational entity responsible for carrying out the act should not be responsible for assuring the act. Also, the "quality" of how servers are managed and ensured, along with how the service quality is handled, should be measured and submitted to a central source database. Often, there is a mentality to "keep the lights on." However, just because the lights are on does not then equate to how effective and efficient the system is. If it is up to the customer to communicate back to the department that implemented it, then there is a major problem in how the organization values and understands "quality." They are viewing the system with a "lights on" or "lights off" mentality.

When I was working at a very large telecommunications company, security, performance, monitoring, and controlling were so crucial that they needed to be accounted for down to the nanosecond. Not because of regulation but because of the company standards and quality. When I was working for a start-up organization that maintained all our webservers in an outsourced Application Service Environment, we had to ensure one hundred percent uptime, and any downtime needed to be communicated. When I worked for a large Fortune 500 institution, things went up and down and only "altered" via monitors. No real justification or control was done. Just simply: "Tell me when the lights are turned off." Different companies have different ideologies of how they handle crucial issues. Some love putting their fireman's hats on and work weekends just to have war stories for raises or job security, while others, when the systems are down, have to justify why there are major issues.

The monitoring, controlling, and security and performance assurance need to be conducted by a separate department to ensure its metrics of the department's ability to do it. I have written an article in the past on ITIL. This is how the assurance of applications can be brought together with the Software Development Life Cycle. They are all part of the ALM and should not be treated separately under "Quality Assurance."

Below is a diagram that describes this.

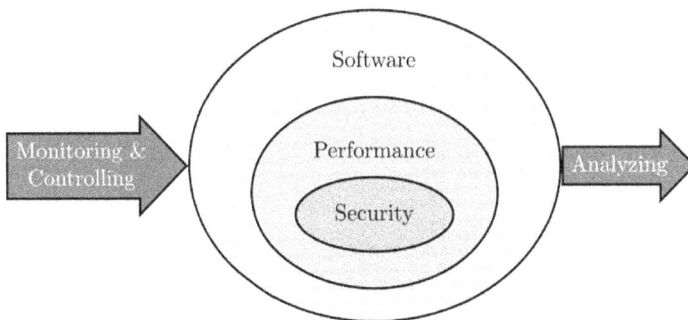

AUTOMATING EMBEDDED SYSTEMS: A GREAT EXAMPLE THAT IT CAN BE DONE!

Is it possible to execute automated testing on a Windows Mobile-embedded Portable Data Terminal (PDT) device? The answer is "yes," thanks to a new solution that I "jimmy rigged."

A PDT is known in the retail space as a "PDT gun," and is sometimes called an Enterprise Digital Assistant (EDA). We've all seen them: ray guns or PDA-type gadgets snugly holstered to the hips of salespeople who work at big box stores. It is a wireless mobile device that scans UPC bar codes for inventory control in retail, warehouse, and hospital settings to name a few. They are also used for onsite management in the contracting and mining industries. The PDT system interacts with the database from the software application where it is used. It is hosted on a traditional server, in a mid-tier or mainframe environment.

I am currently employed at the corporate headquarters of a Fortune 500 retail chain located in Pittsburgh, PA. Not long ago, I overheard a colleague say that it is not possible to use traditional automation tools and methods to create and record automation scripts on PDT/EDA devices. While I shouldn't have been eavesdropping, my coworker's comment got me thinking that there must be some way to make it work.

PDT systems are used across the globe by countless large and small organizations. Having a way to test automate these devices would ease the stress of testing multiple permutation scenarios against multiple PDT systems (e.g., Point of Sale and Inventory Control). Test automation could then be used to add data within these systems via PDT for a test, development, or production environment.

During my quest to find a solution, I scoured the software testing tool industry. I researched every traditional and non-traditional vendor and tool method I could find for test automating on a Windows Mobile-embedded PDT device. I reviewed my company's traditional automated tool vendor, which we use to test automate common client-server and web-based systems, to see if they can interact with a Windows Mobile-embedded system. Since most vendors use a third-party add-on for tools, and vendors provide a solution for mobile test automation, I reviewed those, too.

Surely, one of these tools or vendors would have a solution that addresses this need?

But the answer I eventually found was: "No."

No traditional test automation tool, nor any third-party tool add-on, has scripted over an embedded Windows Mobile-embedded for PDT devices out-of-the-box. One vendor's sales representative told me that they could set up an environment for around $60,000 but could not promise that it would work since it had not been researched or tested yet.

Unable to accept defeat, I started to research less popular, uncommon tools. One was a tool I have used in the past—a limitless product called "EggPlant" by TestPlant (now owned by Keysight Technologies). EggPlant is a UK-based company that uses GUI image recognition to create test automation; it also allows data input and pickup.

EggPlant is built with a two-tiered approach consisting of a controller machine, where scripts are created and executed, and a system under test (SUT) which runs a Virtual Network Computing (VNC) server. Virtual Network Computing (VNC) is a graphical desktop-sharing method that uses the Remote Frame Buffer protocol (RFB) to remotely manage another computer—or in this case, a PDT device—using RealVNC. Another UK-based company, RealVNC provides server and client application for the Virtual Network Computing (VNC) protocol.

EggPlant connects to the VNC server and SUT—which is a PDT device with built-in TCP/IP viewers. The SUT can be any system that has a VNC server installed. In my company's case, the PDT device already had the VNC server installed.

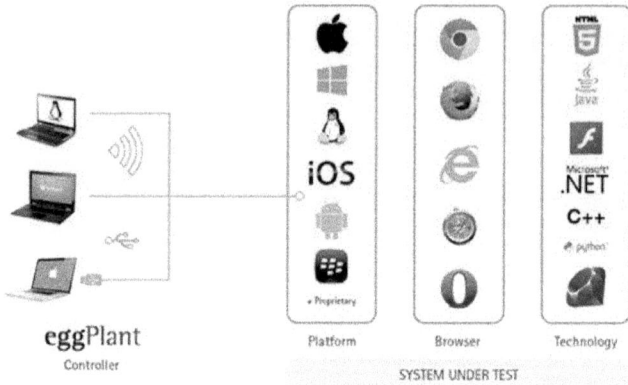

eggPlant
Controller

Platform   Browser   Technology

SYSTEM UNDER TEST

Because of this VNC interaction, EggPlant was the first automation tool on Apple OSX. EggPlant uses SenseTalk scripting, which is an English-like language that is easy to use. SenseTalk is used in conjunction with a "guided record" mode, meaning that the automator instructs EggPlant to direct a system and verify a set of test steps. Additionally, SenseTalk derived from the HyperTalk language used in HyperCard, where it was used for application programming for Apple Macintosh and Apple IIGS systems.

The installation of the RealVNC server on the PDT device had never been done before, and it was a little tricky.

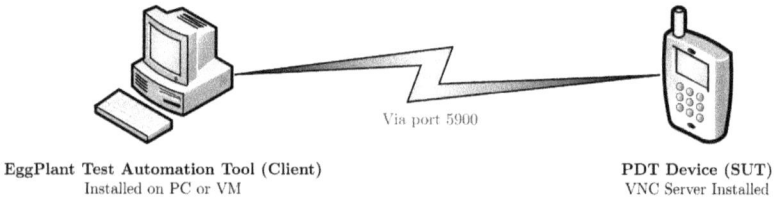

Via port 5900

EggPlant Test Automation Tool (Client)
Installed on PC or VM

PDT Device (SUT)
VNC Server Installed

The PDT device used at my organization was Motorola MC75A, with Avalanche software. A wireless connection to the PC was facilitated by installing Avalanche. I downloaded EggPlant for Windows Mobile and installed files in the directory containing Avalanche on the PC so that the VNC Mobile Server could be executed on the Windows Mobile PDT device.

After the VNC Server was added, I selected and configured options on the VNC Server on the PDT device based on my company's setup.

Just a word of caution: if you choose to use EggPlant as described here, it may take some trial and error until you can make it compatible with your environment setup. The PC will be able to move the PDT device to automate. Make sure that the PDT device is on the network when you connect and make the call from EggPlant to the PDT device via RealVNC to begin the automation. Remember, the purpose of automation is to reduce the burden on the test automator so that he can test other systems. Once your tests are running, you can start multitasking.

When the setup and installation were complete within my organization, and after several configuration changes, we were finally able to connect our PC to a PDT device and create automation. We created a few automation scripts to demo the solution and show its ability to automate on a PDT device. From that, my company attempted a proof of concept (PoC) using EggPlant and this new testing technique. The PoC was conducted over three months by a summer intern. The purpose of the PoC was to see if efficiency could be developed by automating PDT test scenarios. After the PoC, the intern created four EggPlant automation scripts for the PDT.

Later, the intern remarked about how easy the tool was for him to pick up and use—even without prior knowledge of quality assurance and test automation.

It was exciting to discover a solution for test automation on a Windows Mobile-embedded PDT device (specifically a Motorola MC75A), especially since this solution has the potential to impact many areas within the industry. Organizations will realize the true benefit of automation as they increase their efficiency by reducing functional testing time and increase their ability to automate or functionally test more permutations at a faster rate.

I hope that this article improved your understanding of how to achieve test automation for a Windows Mobile embedded PDT / EDA device within your work or industry environment. For more information about EggPlant, visit Keysight.

CHOOSING A MOBILE SOFTWARE TESTING TOOL: WHAT WORKS BEST FOR YOUR NEEDS?

If you are looking for a top-notch mobile testing software to meet your personal or organizational mobile automation needs, there's a good chance you already have a headache with the process. It's easy to become overwhelmed with so many mobile testing tools on the market. In this article, I am pleased to give you advice on how to choose a mobile testing tool based on my many years of experience in the software testing industry.

To start, ask yourself a few questions:

✦ Do I want to integrate with third-party testing tools like Selenium or QTP/UFT?

✦ Will I need to conduct load testing?

✦ Do I have a mobile lab or need access to a mobile lab or emulation services?

When it comes to software testing automation, many companies use Selenium or QTP/UFT. Perfecto Mobile tends to be the easiest to integrate with this automation suite from HP. Another popular mobile tool is Keynote Device Anywhere. Perfecto Mobile and Keynote Device Anywhere offer similar services and solutions, but they are still unique tools and will respond to processes differently. Before choosing which tool is best for you or your company, consider the pros and cons of each by conducting a "proof of concept" PoC exercise.

A PoC needs to ensure the following: Does the tool meet your company processes; does it integrate with current tools; does it meet functional expectations; does it fit your ability and know-how in using, supporting, and maintaining; lastly, does it fit your price and yearly cost for support? Second, you need to look at a potential internal project to test the tool on. Look at a project where it can be best utilized. For example, one of my previous companies needed a load-testing mobile and web solution to help fix a short-term need for a production problem identification. I used two tools and showed how fast each can give a solution to the issue and how well it can be implemented and display the issue. I then took the results and put them on a PowerPoint slide so that the company technology leaders and department leaders could better understand the results and choose a solution. In this case, the mobile and web PoC was an example where I selected Tricentis NeoLoad.

The best tool that I know of, which solely loads test mobile platforms, is Tricentis NeoLoad. In fact, you can also create simulated networks and wireless devices and run them at the same time over a variety of networks, including different PC or Mac environments on different browsers. Tricentis NeoLoad assists with load testing mobile, wireless, and carrier services to ensure load scalability.

Again, when looking to automate mobile testing, the best thing to do is to conduct a proof of concept and completely integrate your organization's testing process into the selected tool. Technology, processes, solutions, and accessibility change over time, so it's best to slowly and comprehensively integrate such tools into your projects.

THE ADVANCEMENTS OF ARTIFICIAL INTELLIGENCE IN SOFTWARE TESTING AND PROJECT MANAGEMENT

Artificial Intelligence (AI) is increasingly being integrated into software development processes, influencing how software is built, tested, and delivered. Applications such as predictive analytics and autonomous testing are changing established workflows and affecting roles throughout the software development lifecycle. This chapter examines the impact of AI on software testing and project management, along with the related opportunities and challenges.

AI is changing software testing by moving from manual testing to automated processes. Key impacts include: Test Case Generation, where AI algorithms examine requirements and user stories to generate test cases; Self-Healing Tests, where AI-powered frameworks detect UI or API changes and update test scripts; Predictive Defect Analysis, where machine learning models identify code areas likely to have defects using historical data; Intelligent Test Prioritization, where AI ranks test cases based on risk, usage patterns, and code changes; and Natural Language Processing (NLP), which allows testers to describe scenarios in natural language so AI can convert them into executable scripts for improved communication between business and technical teams.

AI is transforming project management, enabling smarter decisions and faster, higher-quality delivery. Instead of relying on traditional methods, managers now use AI for: precise effort estimation, early risk prediction, automated reporting, team sentiment analysis, and optimized scheduling.

There are several challenges and ethical considerations when using AI in Project Management. Although AI can provide significant advantages, it also presents specific complexities: Bias in Algorithms, AI models trained on unbalanced data may perpetuate biases in testing and decision-making. Transparency, some models function as black boxes, making it challenging to explain certain decisions or predictions. Skill Gaps, Teams may need to develop new skills to use AI tools effectively and interpret their results. Overreliance, Excessive dependence on AI systems may result in missed edge cases or overlooked human factors.

It is important to view the future of work as a collaborative effort between artificial intelligence (AI) and professionals, rather than a process of replacement. AI serves to enhance the capabilities of testers and project managers by augmenting human creativity and judgment with increased accuracy and efficiency. New roles, such as "AI Test Strategist" and "AI Project Analyst," are likely to emerge in hybrid teams. Additionally, AI systems will continually adapt to projects, learning from each sprint and release cycle. Despite these advancements, human oversight will remain essential for critical decisions, maintaining ethical standards and contextual appropriateness.

In summary, successfully navigating the evolving AI-driven landscape is vital. The incorporation of artificial intelligence into software testing and project management signifies a transformative shift rather than a temporary trend. Organizations embracing these innovations can realize faster delivery, heightened quality, and stronger collaboration. Nevertheless, achieving success requires not only technological implementation but also a cultural shift, adherence to ethical principles, and a commitment to continuous learning.

As we advance, the question is no longer whether artificial intelligence will transform software development; rather, it is how we will guide this transformation to effectively meet the objectives of our organizations, teams, and users.

CONCLUSION

I hope that these tips and strategies help to enhance your management towards better quality assurance within your company and projects. Remember that culture, quality, and perception are reality until proven otherwise. It takes your efforts to prove the "otherwise" to ensure stronger quality for the product, end users, and the customers.

In the ever-evolving landscape of IT Software technology and business, the integration of strong project management practices with severe software quality assurance (QA) methodologies is vital. This book has explored the crucial intersections between these two domains, emphasizing the importance of a holistic approach to delivering high-quality software and IT products.

Successful project management delivers the framework for planning, testing, executing, and closing projects successfully. It confirms that projects are completed on time, within budget, and to the happiness of all stakeholders. On the other hand, software and IT quality assurance aims to prevent defects and ensure that the final product meets or exceeds customer expectations. Jointly, they form a powerful collaboration that drives project success and product superiority.

Throughout this book, we have examined numerous strategies, tools, and best practices that can help project managers and QA professionals cooperate more effectively. From defining clear quality standards and establishing comprehensive QA plans to leveraging advanced project management.

As we conclude, it is essential to recognize that both project management and software quality assurance are endless processes. They require ongoing commitment, alteration, and improvement. By fostering a culture of quality and collaboration, organizations can not only meet their abrupt project goals but also build a foundation for long-term success and innovation.

In summary, the journey of integrating project management with software and IT quality assurance is one of continuous learning and improvement within the company culture. By embracing these tips and strategies discussed in this book, you are well on your way to achieving excellence in your projects and delivering software that truly stands out in the market.

I have participated in numerous computer associations; however, the International Institute for Software Testing (IIST) distinguishes itself as a leader in the field (testinginstitute.com). Established in 1999, IIST is recognized as the world's largest provider of software testing training and is dedicated to advancing professionalism through education-based certifications and conferences focused on software testing and quality assurance. Involvement by Project Managers, CIOs, and CTOs in these programs has proven beneficial to organizations.

A notable development from this organization is Rommana Software, created by Dr. Magdy Hanna (rommanasoftware.com). This robust ALM tool seamlessly integrates project management, test management, document management, requirements and user story management, issue management, change management, collaboration management, and use case management into a single system. For those seeking a comprehensive solution that interconnects these critical functions, Rommana Software merits consideration.

CLOSING ADVICE

Often, people assume where they should be in their careers and may even push themselves into a position when their abilities are naturally stronger than what they had realized. Aim for excellence, while recognizing that others may not always be striving for their highest potential. Your capabilities may exceed your own expectations. It is essential to remain humble and learn from mistakes, as this promotes development and insight. Letting ego influence your actions can impede both personal growth and your ability to contribute positively to the organization and advance your career.

BIBLIOGRAPHY

Yasar, Kinza. "What is cloud computing? Types, examples and benefits." TechTarget, June 2024. http://searchenterprisedesktop.techtarget.com/sDefinition/0,,sid192_gci1287881,00.html.

Beck, Melinda. "Manner: Advocating for a Relative in the Hospital." The Wall Street Journal, Oct. 28, 2008.

www.ingramcontent.com/pod-product-compliance
Lightning Source LLC
Chambersburg PA
CBHW040903210326
41597CB00029B/4944